REACTIVITY *and* TRANSPORT *of* HEAVY METALS *in* SOILS

REACTIVITY *and* TRANSPORT *of* HEAVY METALS *in* SOILS

H. Madgi Selim
Louisiana State University
Baton Rouge, Louisiana

Michael C. Amacher
U.S. Department of Agriculture Forest Service
Logan, Utah

LEWIS PUBLISHERS

Boca Raton   New York   London   Tokyo

Acquiring Editor	Joel Stein
Project Editor	Carole Sweatman
Marketing Manager	Greg Daurelle
Cover Design	Denise Craig
Manufacturing	Sheri Schwartz

Library of Congress Cataloging-in-Publication Data

Selim, Hussein Magdi Eldin, 1944–
    Reactivity and transport of heavy metals in soils / by H. M. Selim
and M. C. Amacher.
        p. cm.
    Includes bibliographical references and index.
    ISBN 0-87371-473-3 (alk. paper)
    1. Soils—Heavy metal content.   2. Heavy metals—Environmental
aspects.   3. Soils—Solute movement.   I. Amacher, Michael C.
II. Title.
S592.6.H43S45   1996
628.5′5—dc20   96-11372
                                                                            CIP

© 1997 by CRC Press, Inc.
Lewis Publishers is an imprint of CRC Press

No claim to original U.S. Government works
International Standard Book Number 0-87371-473-3
Library of Congress Card Number 96-11372

To our wives

Liz and Jan

Preface

The subject of transport of solutes in porous media is one of the fascinating areas of science. The fate of many heavy metal solutes in the environment is of some concern because of their potential reactivity, toxicity, and mobility in the soil. Moreover, heavy metals in the soil undergo a series of complex interactions that are governed by soil properties and conditions. Understanding of such complex interactions is a prerequisite in the effort to predict the behavior and mobility of heavy metals in the soil system.

Although the reactivity and transport of heavy metals in soils are the subject of this book, the modeling approaches described herein are applicable to other reactive solutes in soils. Most of the modeling applications (examples) presented in this book deal with Cr(VI), Cd, and Zn because of their economic importance and potential toxicity. Other heavy metals used in modeling applications include Hg, Pb, and Cu.

Due to the presence of elevated levels of heavy metals, many sites have been identified as hazardous waste sites. Sources of contamination include industrial waste by-products and abandoned mines. In some cases, contamination of surface-water streams and groundwater with metals of potential health effects have also been identified. The soil environmental conditions (physical, chemical, and biological processes) are significant factors affecting the fate of heavy metals in soils.

This book should be useful to scientists, engineers, and undergraduate and graduate students in soil science, environmental science and engineering, chemical and civil engineering, hydrology and geology. We hope that it is useful for federal and state agencies, researchers, consulting engineers, and decision makers in the management and cleanup of metal contaminated sites.

The book is organized in eight chapters. In Chapter 1, the use and limitation of equilibrium models for describing metal transport in soils are developed and discussed. Because of the limitations of equilibrium retention models for describing metal reactions and transport in soils, most of the remainder of the subsequent chapters are devoted to describing the development and applications of kinetic retention approaches. Laboratory methods of obtaining kinetic data, data analysis, and model evaluation/validation is addressed in Chapter 2. In subsequent chapters (4 to 7), we present principles of solute transport and mathematical models as a tool for the description of the retention reactions of heavy metals during transport in the soil.

In Chapter 3, we discuss single and multiple retention reactions in soils with emphasis on commonly observed nonlinear behavior of heavy metals in soils. These reactions are then incorporated into the convection–dispersion transport equation in Chapter 4, and the appropriate selection of boundary and initial condition are discussed. In Chapter 5, we present second-order two-site modeling approaches with application to several examples of Cr(VI) transport in soils. In Chapter 6, physical nonequilibrium approaches based on multiregion (mobile–immobile) concepts are presented. In Chapter 7, we discuss multicomponent approaches with special emphasis on competitive ion exchange and specific sorption mechanisms during transport. Finally, mechanisms of heavy metals in soils that are consistent with the modeling approaches described in earlier chapters are presented in Chapter 8.

The authors wish to thank various individuals for their assistance in completing this book. Special thanks goes to our colleagues who took the time to review and provide valuable suggestions on this book; Phil M. Jardine, Oak Ridge National Laboratory (Chapter 3); Charles W. Boast, University of Illinois (Chapter 4); Liwang Ma, Louisiana State University (Chapter 5); Robert S. Mansell, University of Florida (Chapter 6); Lewis A. Gaston, USDA-ARS (Chapter 7); and Iskandar K. Iskandar, the Cold Regions Research and Engineering Laboratory. We also wish to acknowledge our current and former graduate students and other individuals who contributed directly and indirectly to this book; Davis S. Burden, Bar Y. Davidoff, Bernhard Buchter, Christoph Hinz, Ezio Buselli, Shikui K. Xue, Susan Luque, and Hongxia Zhu. Finally, we are grateful to our families for their love and support during the preparation of this book. We dedicate this book to our wives.

H. Magdi Selim
Baton Rouge, Louisiana

Michael C. Amacher
Logan, Utah

The Authors

H. Magdi Selim is Professor of Soil Physics at Louisiana State University, Baton Rouge, Louisiana. Dr. Selim received his M.S. and Ph.D. in soil physics from Iowa State University, Ames, Iowa in 1969 and 1971, respectively, and his B.S. in soil science from Alexandria University, Alexandria, Egypt in 1964. Dr. Selim has published more than 100 papers and book chapters and is a co-editor of one book. His research interests concern modeling the mobility of contaminants and their retention behavior in soils and groundwaters. His research interests also include saturated and unsaturated water flow in multilayered soils. He received the LSU Chapter—Phi Kappa Phi Young Faculty Award in 1980, the First Mississippi Research Award for outstanding achievements in the Louisiana Agricultural Experiment Station in 1982, and Gamma Sigma Delta Outstanding Research Award in 1991. He served as an associate editor of *Water Resources Research* and *Soil Science Society of America Journal*. Dr. Selim is a Fellow at the American Society of Agronomy and the Soil Science Society of America.

Michael C. Amacher is Research Soil Chemist at the Forestry Sciences Laboratory, Intermountain Research Station, U.S. Department of Agriculture—Forest Service, Logan, Utah. Dr. Amacher received his B.S. and M.S. degrees in chemistry from Pennsylvania State University in 1972 and 1978, respectively, and his Ph.D. degree in soil chemistry in 1981, also from Pennsylvania State University. Dr. Amacher has authored or co-authored over 50 publications. His research interests include modeling the reactions and transport of elements in soils and aquatic environments and developing methods to remediate and restore watersheds disturbed by mining and other resource use activities. He served as an associate editor of *Soil Science Society of America Journal*.

Contents

Preface

The Authors

1. Introduction 1

    Overview 1

    Equilibrium Retention Models 4
        Freundlich 4
        Langmuir 10
        Two-Site Langmuir 11

    General Isotherm Equations 11

    Asymptotic Behavior 14

2. Methods of Obtaining and Analyzing Kinetic Data 17

    Basic Kinetic Concepts 17

    Types of Reactions in Soil Environments 20

    Methods of Obtaining Kinetic Data 22
        Relaxation Methods 22
        Batch Methods 22
            Batch Reactor Design 23
            Mixing Techniques 26
            Separation of Solid and Liquid Phases 29
            In Situ Methods 30
            Radiotracer Methods 30
            Advantages and Limitations 32
        Flow Methods 34
            Thin Disk Method 35
            Stirred-Flow Method 36
            Fluidized Bed Reactor 38
            Column Method with Batch Control 38
            Radiotracer Methods 38
            Advantages and Limitations 39
            Comparison of Batch and Flow Methods 39
        Special Methods for Studying Desorption Reactions 40
            Flow Methods 40

Dilution Methods 40
Infinite Sink Methods 42
Special Methods for Separating Reactions 42

Data Analysis Methods 43
Initial Rate Method 44
Method of Isolation 44
Graphical Methods 45
Rate Coefficient Constancy 45
Fractional Lives Methods 46
Parameter-Optimization Methods 46

Effects of Experimental Variables on Rate Functions 48
Reactant Concentrations 48
Temperature 48
pH 49
Ionic Strength 49
Solution Composition 49

Column Transport Methods 49

3. **Kinetic Retention Approaches** 55

Single-Reaction Models 56

Multiple-Reaction Models 61
Two-Site Models 62
Multireaction Models 63

Sensitivity Analysis 65

Applications 67
Estimation of Overall Reaction Order 67
Variations of MRM 69
Description of Batch Kinetic Data with MRM 69

One Site or Two Sites 73

4. **Transport** 77

Continuity Equation 77

Transport Mechanism 80

Steady-State Flow 83

Boundary and Initial Conditions 84

Exact Solutions 85

Brenner Solution 86
Lindstrom Solution 87
Cleary and Adrian Solution 87

Estimation of D 91

Other Exact Solutions 94

Nonlinear Retention and Transport 94

Numerical Solution 97

Sensitivity Analysis 100

Applications 107

5. A Second-Order, Two-Site Transport Model 113

Second-Order Kinetics 113

Transport Model 117

Sensitivity Analysis 119
Kinetics 119
Transport 120

Applications 127
Estimation of S_{max} and F 127
Reaction Kinetics 128
Transport 131

6. Mobile–Immobile or Two-Region Transport Approaches 135

General Information 135

Estimation of α 138

Estimation of θ^m and θ^{im} 140

A Second-Order Approach 141
Initial and Boundary Conditions 145

Sensitivity Analysis 146

Applications 149

A Modified Two-Region Approach 152

7. Multicomponent and Competitive Ion Exchange Approaches 157

     General Assumptions 158

     Equilibrium Ion Exchange 159

     Binary Homovalent Systems 160

     Multiple Ion Systems 161

     Variable Selectivities 163

     Kinetic Ion Exchange 164

     Specific Sorption 166

     Simulations 167

     Applications 170

8. Relationship between Models and Soil Chemical Reactions 175

     Ion Exchange and Surface Complexation Kinetics 175

     Surface Complexation Equilibrium 178

     Model Selection 180

     Irreversible Mechanisms 181

References 183

Index 197

1 INTRODUCTION

Heavy metals in the environment are a source of some concern because of their potential reactivity, toxicity, and mobility in the soil. Some heavy metals (e.g., Cu and Zn) are essential for plant and animal health. However, at environmental concentrations above those necessary to sustain life, toxicity may occur. Other heavy metals (e.g., Cd and Pb) are not known to be essential to plants and animals. Toxicity may occur when these metals become concentrated in the environment above background levels.

Sources of heavy metals in the environment and factors influencing their distribution, reactivity, mobility, and toxicity are numerous. Several reviews on metals in the environment are available and the reader may refer to the work of Förster and Wittman (1981) and Adriano (1986), among others. In this book, we focus on describing the transport of heavy metals in the soil environment with emphasis on the reactivity of metals with soil surfaces during transport processes. Although numerous literature citations are included, this book is not a literature review. Rather, we focus on presentation of various approaches to modeling the *Reactivity and Transport of Heavy Metals in Soils*. Moreover, the modeling approaches described herein are applicable to other reactive solutes present in the soil. Most of the modeling examples focus on Cr(VI), Cd, and Zn because of economic importance and potential toxicity. Other heavy metals used in modeling applications include Hg, Pb, and Cu.

OVERVIEW

Heavy metals in the soil environment can be involved in a series of complex chemical and biological interactions. Several of the reactions include oxidation–reduction, precipitation–dissolution, volatilization, and surface–solution phase complexation. A number of scientists have studied soil properties that significantly affect the behavior of heavy metals such as cadmium in the soil (e.g., Kabata-Pendias and Pendias, 1984; Buchter et al. 1989). The problem of identifying the fate of heavy metals in soils must account for retention reactions and transport of the various species in the soil environment (Theis, 1988; Barrow, 1989).

Retention reactions in soils are important processes governing the fate of chemical contaminants such as heavy metals in groundwaters. The ability to predict the mobility of heavy metals in the soil and the potential contamination of groundwater supplies is a prerequisite in the management of land disposal of chemical contaminants. Mathematical models that describe the potential mobility of heavy metals must include a description of the retention processes in the soil matrix. Concern about land and water quality has led to an increased interest in the understanding of the processes of retention of contaminants and their potential mobility in the soil profile.

To predict the transport of heavy metals, models that include retention and release reactions of solutes with the soil matrix are needed. Retention and release reactions in soils include precipitation–dissolution, ion exchange, and adsorption–desorption reactions (Amacher et al., 1986). Retention and release are influenced by a number of soil properties including texture, bulk density, pH, organic matter, and the type and amount of clay minerals. Adsorption is the process whereby solutes bind to surfaces of soil particles to form outer- or inner-sphere solute–surface site complexes; whereas, ion exchange is the process whereby charged solutes replace ions on soil particles. Adsorption and ion exchange are related in that an ionic solute species may form a surface complex and may replace another ionic solute species already on the surface binding site. Strictly speaking, the term "retention" or the commonly used term "sorption" should be used when the mechanism of solute removal from solution by soil is not known, and the term "adsorption" should be used only to describe the formation of solute-surface site complexes. However, the term "adsorption" is often used to include all processes mentioned above, even though the processes cannot be distinguished in most experiments.

Extensive research has been carried out for the purpose of describing the retention/release behavior of several heavy metals in soils. Fuller (1977), Alesii et al. (1980), Dowdy and Volk (1983), Ellis et al. (1983), and Kabata-Pendias and Pendias (1984), among others, presented an overview of retention–release and leaching investigations for several heavy metals in soils. These investigations also describe the physical and chemical properties of soil that influence the fate of heavy metals in the soil environment and their potential leaching to groundwater supplies. However, over the last two decades, only a limited number of investigations attempted to quantify the mobility of heavy metals in the soil profile. Specifically, mathematical models that describe the transport of heavy metals in laboratory soil columns or in soil profiles under field conditions have only been introduced recently in the literature.

Sidle et al. (1977) were among the earliest researchers to utilize the convection–dispersion equation for the description of Cu, Zn, and Cd movement in a sludge-treated forest soil. The primary feature of their model is that the retention/release mechanism was assumed to be fully reversible and of the nonlinear equilibrium (Freundlich) type. Model calculations resulted in underprediction of the mobility of these metals at two depths. A similar approach was used by Amoozegar-Fard et al. (1983) and van Genuchten and Wierenga (1986), where a linear equilibrium sorption mechanism was incorporated into the convection–dispersion equation to describe Cr(VI) mobility in soil columns. Schmidt and Sticher (1986) found that the equilibrium retention of Cd, Pb, and Cu was successfully described by a two-site sigmoidal Langmuir isotherm equation.

For several heavy metals (e.g., Cr, Cu, Zn, Cd, and Hg), retention–release reactions in the soil solution were observed to be strongly time dependent. Studies on the kinetic behavior of the fate of several heavy metals in soils included those of Harter (1984), Aringhieri et al. (1985), and Amacher et al. (1986), among others. A number of empirical models were proposed to describe kinetic retention/release reactions of solutes in the solution phase. The earliest model was the first-order kinetic equation, first incorporated into the convection–dispersion transport equation by Lapidus and Amundson (1952). First-order kinetic reactions have been extended to include the nonlinear kinetic type (van Genuchten et al., 1974; Mansell et al., 1977; Fiskell et al., 1979). A variety of other kinetic reactions were given by Murali and Aylmore (1983). Amacher et al. (1986) found that the use of single-reaction kinetic models did not adequately describe the time-dependent retention of Cr, Hg, and Cd for several initial concentrations and several soils. As a result, Amacher et al. (1988) developed a multireaction model that included concurrent and concurrent-consecutive processes of the nonlinear kinetic type. The model was capable of describing the retention behavior of Cd and Cr(VI) with time for several soils. In addition, the model predicted that a fraction of these heavy metals was irreversibly retained by the soil. A literature search revealed that no studies were carried out on the description of heavy metal transport in soils where the retention–release reactions were based on kinetic mechanisms. The study of Amoozegar-Fard et al. (1984) was perhaps the first study to investigate the mobility of Cd, Ni, and Zn using a fully reversible first-order kinetic reaction.

The fate of heavy metals in soils depends on retention reactions and transport of the various species in the soil. In fact, Barrow (1989) stated that the use of single-reaction models is not adequate since such models describe the fate of only one species, with no consideration of the simultaneous reactions of others in the soil system. This is supported by the work of Amacher et al. (1986) who showed that sorption–desorption of Cd, Cr, and Hg from batch studies on several soils was not described using a single reaction model of the equilibrium Langmuir or Freundlich type. Models that consider several processes—including ion exchange, complexation, dissolution/precipitation, and competitive adsorption—are FIESTA (Jennings et al., 1982), CHEMTRAN (Miller and Benson, 1983), and TRANQL (Cederberg et al., 1985), among others. A major disadvantage of multicomponent models lies in the basic assumption of local equilibrium of the governing reactions. In addition, due to their complexity, several of these models have not been fully validated. Kirkner et al. (1985) utilized the FIESTA model to describe Ni and Cd breakthrough results on a sandy soil. Model predictions provided higher retardation of Cd and lower retardation for Ni. However, improved predictions were obtained when a kinetic approach was used with approximate parameters obtained from batch experiments.

Alternatives to multicomponent models are the simpler multisite or multireaction models that deal with the multiple interactions of one species in the soil environment. Such models are empirical in nature and are based on the assumption that a fraction of the total sites (type 1 sites) is time-dependent (i.e., kinetic in nature), whereas the remaining fraction (type 2 sites) interacts rapidly or instantaneously with solute in the soil solution (Selim et al., 1976; Jardine et al., 1985). Nonlinear equilibrium (Freundlich) and first- or nth-order kinetic reactions are the associated processes. Such a two-site approach proved successful in describing observed extensive tailing of breakthrough results (for a review, see Selim et al., 1990). Another two-site approach was proposed by Theis et al. (1988) for Cd mobility and adsorption on goethite. They assumed that the nature of reactions for both sites was governed by second-order kinetic reactions. The reactions were assumed to be consecutive, where the second reaction was irreversible in nature. Amacher et al. (1988) developed a multireaction model that included concurrent and concurrent-consecutive processes of the nonlinear kinetic type. The model was capable of describing the retention behavior of Cd and Cr(VI) with time for several soils. In addition, the model predicted that a fraction of these heavy metals was irreversibly retained by the soil. Amacher et al. (1990) concluded that the multireaction model was also successful in describing adsorption of Hg for several soils.

In the next section, the use and limitation of equilibrium models for describing metal transport in soils are developed and discussed. Because of the limitations of equilibrium retention models for describing metal reactions and transport in soils, most of the remainder of the subsequent chapters is devoted to describing the development and applications of kinetic retention approaches.

EQUILIBRIUM RETENTION MODELS

The form of solute retention reactions in the soil system must be identified if prediction of the fate of reactive solutes such as heavy metals in the soil is sought. Sorption or exchange has been described by either instantaneous equilibrium or a kinetic reaction where concentrations in solution and sorbed phases vary with time. Reviews of various forms of equilibrium and kinetic models are given by Murali and Aylmore (1983), Selim (1989), and Selim et al. (1990). Nielsen et al. (1986) presented a comprehensive discussion of significant features of sorption-exchange reactions of the equilibrium and kinetic type. Linear, Freundlich, and one- and two-site Langmuir equations are perhaps the most commonly used to describe equilibrium reactions. In the subsequent sections we discuss Freundlich and Langmuir reactions and their use in describing equilibrium retention. In subsequent chapters, we will discuss kinetic type reactions and their implication for single and multireaction retention and transport models.

Freundlich

The Freundlich equation is perhaps the simplest approach for quantifying the behavior of retention of reactive solute with the soil matrix. It is certainly one of the oldest of the nonlinear sorption equations and has been used widely to describe solute retention by soils (Helfferich, 1962; Sposito, 1984; Travis and Etnier, 1981; Murali and Aylmore, 1983). The Freundlich equation is

$$S = K_f C^b \qquad (1\text{-}1)$$

where S is the amount of solute retained by the soil, in $\mu g\ g^{-1}$ or mg kg^{-1}; C is the solute concentration in solution in mg L^{-1} or μg mL^{-1}; K_f is the distribution coefficient in L kg^{-1} or mL g^{-1}; and the parameter b is dimensionless and typically has a value of $b < 1$. The distribution coefficient describes the partitioning of a solute species between solid and liquid phases over the concentration range of interest and is analogous to the equilibrium constant for a chemical reaction. For $b = 1$, the Freundlich equation is often referred to as the linear retention equation:

$$S = K_d C \qquad (1\text{-}2)$$

where K_d is the linear distribution coefficient (mL g^{-1}), which is commonly referred to in the literature. Selected equilibrium and kinetic-type retention models for heavy metal retention in soils are given in Table 1-1. Kinetic-type models will be discussed in Chapter 3.

Although the Freundlich equation has been rigorously derived (Sposito, 1980), the goodness-of-fit of the Freundlich equation to solute retention data does not provide definitive information about the actual processes involved, since the equation is capable of describing data irrespective of the actual retention mechanism. Often complex retention processes can at least in part be described

TABLE 1-1 Selected Equilibrium and Kinetic-Type Models for Heavy Metal Retention in Soils

Model	Formulation[a]
Equilibrium type	
Linear	$S = K_d C$
Freundlich	$S = K_f C^b$
General Freundlich	$S/S_{max} = [\omega C/(1 + \omega C)]^\beta$
Rothmund-Kornfeld ion exchange	$S_i/S_T = K_{RK}(C_i/C_T)^n$
Langmuir	$S/S_{max} = \omega C/[1 + \omega C]$
General Langmuir-Freundlich	$S/S_{max} = (\omega C)^\beta/[1 + (\omega C)^\beta]$
Langmuir with sigmoidicity	$S/S_{max} = \omega C/[1 + \omega C + \sigma/C]$
Kinetic type	
First order	$\partial S/\partial t = k_f(\Theta/\rho)C - k_b S$
n^{th} order	$\partial S/\partial t = k_f(\Theta/\rho)C^n - k_b S$
Irreversible (sink/source)	$\partial S/\partial t = k_s(\Theta/\rho)(C - C_p)$
Second-order irreversible	$\partial S/\partial t = k_s(\Theta/\rho)C(S_{max} - S)$
Langmuir kinetic	$\partial S/\partial t = k_f(\Theta/\rho)C(S_{max} - S) - k_b S$
Elovich	$\partial S/\partial t = A \exp(-BS)$
Power	$\partial S/\partial t = K(\Theta/\rho)C^n S^m$
Mass transfer	$\partial S/\partial t = K(\Theta/\rho)(C - C^*)$

[a] $A, B, b, C^*, C_p, K, K_d, K_{RK}, k_b, k_f, k_s, n, m, S_{max}, \omega, \beta$, and σ are adjustable model parameters, ρ is bulk density, Θ is volumetric soil water content, C_T is total solute concentration, and S_T is total amount sorbed of all competing species.

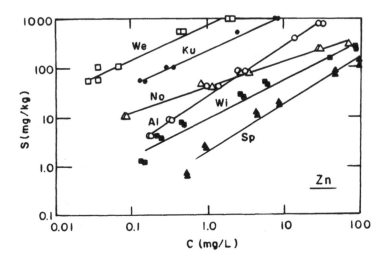

Figure 1-1   Retention isotherms for Zn on selected soils. The soils are represented by
Alligator (Al), Kula (Ku), Norwood (No), Spodosol (Sp),Webster (We), and
Windsor (Wi).

by relatively simple models such as the Freundlich equation. Therefore, the
Freundlich parameters K_f and b are best regarded as descriptive parameters in
the absence of independent evidence concerning the actual retention mechanism.

There are numerous examples for solute retention, as described successfully
by the use of the Freundlich equation (see Sposito, 1984; Travis and Etnier, 1981;
Murali and Aylmore, 1983; Sparks, 1989). Logarithmic representation of the
Freundlich equation is frequently used to represent the data as illustrated in
Figures 1-1 to 1-3. Here, the slope of the best-fit curve provides the nonlinear
parameter b and the intercept as K_f according to [$\log(S) = K_f + b \log(C)$] as

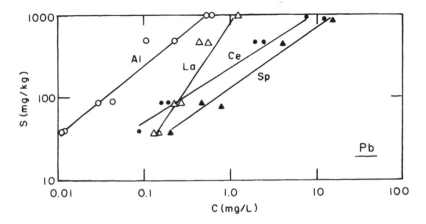

Figure 1-2   Retention isotherms for Pb on selected soils. The soils are represented by
Alligator (Al), Cecil (Ce), Lafitte (La), and Spodosol (Sp).

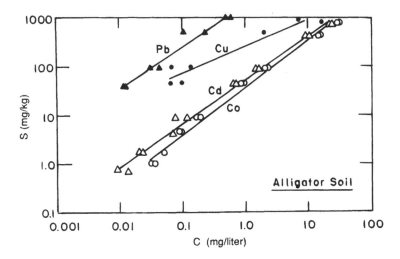

Figure 1-3 Retention isotherms for selected cation species on Alligator soil.

long as a linear representation of the data in the log form is achieved. In Figures 1-1 and 1-2, the use of the Freundlich equation for Zn and Pb retention for several soils is illustrated; whereas, Figure 1-3 shows Pb, Cu, Cd, and Co isotherms for one (Alligator) soil (Buchter et al., 1989).

Buchter et al. (1989) studied the retention of 15 elements by 11 soils from 10 soil orders to determine the effects of element and soil properties on the magnitude of the Freundlich parameters K_f and b. The names, taxonomic classification, and selected properties of the 11 soils used in their study are listed in Table 1-3, and estimated values for K_f and b for selected heavy metals are given in Table 1-2. A wide range of K_f values, from 0.0419 to 4.32×10^7 mL g^{-1}, were obtained, which illustrates the extent of heavy metals affinity among various soil types. Such a wide range of values was not obtained for the exponent parameter b, however. The magnitudes of K_f and b were related to both soil and element properties. Strongly retained elements such as Cu, Hg, Pb, and V had the highest K_f values. The transition metal cations Co and Ni had similar K_f and b values, as did the group IIB elements Zn and Cd. Oxyanion species tended to have lower b values than did cationic species. Soil pH and CEC (cation exchange capacity) were significantly correlated to log K_f values for cationic species. High pH and high CEC soils retained greater quantities of the cationic species than did low pH and low CEC soils. A significant negative correlation between soil pH and the Freundlich parameter b was observed for cationic species, whereas a significant positive correlation between soil pH and b for Cr(VI) was found. Greater quantities of anionic species were retained by soils with high amounts of amorphous iron oxides, aluminum oxides, and amorphous material than were retained by soils with low amounts of these minerals. Several anionic species were not retained by high-pH soils. Despite the facts that element retention by soil is the result of many interacting processes and that many factors influence

TABLE 1-2 Freundlich Model Parameters K_f and b for Selected Soils and Heavy Metals

Element	Initial species	Alligator	Calciorthid	Cecil	Kula	Lafitte	Molakai	Norwood	Olivier	Oldsmar	Webster	Windsor	Mean
						K_f (ml/g)							
Co	Co^{2+}	3.57E+01	2.51E+02	6.56E+00	1.05E+02	3.39E+01	9.25E+01	2.74E+01	6.70E+01	2.55E+00	3.63E+02	6.28E+00	3.75E+01
Ni	Ni^{2+}	3.78E+01	2.06E+02	6.84E+00	1.10E+02	5.01E+01	4.49E+01	2.09E+01	5.05E+01	3.44E+00	3.37E+02	8.43E+00	3.61E+01
Cu	Cu^{2+}	2.58E+02	2.62E+03	5.37E+01	2.05E+03	2.21E+02	3.68E+02	8.91E+01	2.18E+02	5.62E+01	6.35E+03	7.71E+01	3.17E+03
Zn	Zn^{2+}	2.81E+01	4.20E+02	1.12E+01	2.38E+02	2.01E+01	8.04E+01	4.21E+01	8.91E+01	2.12E+00	7.74E+02	9.68E+00	4.79E+01
Cd	Cd^{2+}	5.25E+01	2.88E+02	1.39E+01	1.87E+02	5.27E+01	9.12E+01	2.88E+01	9.79E+01	5.47E+00	7.55E+02	1.44E+01	5.93E+01
Hg	Hg^{2+}	1.08E+02	1.96E+01	8.13E+01	2.49E+02	1.90E+02	1.20E+02	1.13E+02	1.29E+02	8.63E+01	2.99E+02	1.30E+02	1.15E+02
Pb	Pb^{2+}	1.81E+03	—	2.36E+02	4.32E+07	9.18E+02	8.17E+03	3.85E+02	1.64E+04	1.36E+02	—	4.72E+02	3.37E+02
V	VO_3^-	1.42E+02	1.08E+01	3.97E+01	2.22E+03	1.03E+02	5.05E+02	1.86E+01	9.12E+01	9.08E+01	8.07E+01	1.53E+01	1.03E+02
Cr	CrO_4^{2-}	3.41E+00	—	—	6.28E+01	3.03E+01	6.41E+00	—	—	5.47E+00	—	8.47E+00	1.12E+01
Mo	$Mo_7O_{24}^{6-}$	5.75E+01	—	1.80E+01	4.11E+02	8.15E+01	1.18E+02	—	—	2.56E+01	—	4.38E+01	6.44E+01
As	AsO_4^{3-}	4.78E+01	8.87E+00	1.98E+01	1.50E+03	7.10E+01	1.56E+02	8.53E+00	4.60E+01	1.87E+01	2.36E+01	1.05E+02	4.71E+01
						b							
Co	Co^{2+}	0.953	0.546	0.745	0.878	1.009	0.621	0.627	0.584	0.811	0.782	0.741	0.754
Ni	Ni^{2+}	0.939	0.504	0.688	0.738	0.903	0.720	0.661	0.646	0.836	0.748	0.741	0.739
Cu	Cu^{2+}	0.544	1.140	0.546	1.016	0.987	0.516	0.471	0.495	0.602	1.420	0.567	0.755
Zn	Zn^{2+}	1.011	0.510	0.724	0.724	0.891	0.675	0.515	0.625	0.962	0.697	0.792	0.739
Cd	Cd^{2+}	0.902	0.568	0.768	0.721	0.850	0.773	0.668	0.658	0.840	0.569	0.782	0.736
Hg	Hg^{2+}	0.741	0.313	0.564	1.700	0.751	0.960	0.582	1.122	0.513	2.158	0.681	1.008
Pb	Pb^{2+}	0.853	—	0.662	5.385	0.558	1.678	0.741	0.998	0.743	—	0.743	1.485
V	VO_3^-	0.592	0.857	0.629	1.402	0.679	0.847	0.877	0.607	0.483	0.762	0.647	0.762
Cr	CrO_4^{2-}	0.504	—	—	0.609	0.374	0.607	—	—	0.394	—	0.521	0.501
Mo	$Mo_7O_{24}^{6-}$	0.882	—	0.617	1.031	0.607	0.664	—	—	0.451	—	0.544	0.685
As	AsO_4^{3-}	0.636	0.554	0.618	1.462	0.747	0.561	0.510	0.548	0.797	0.648	0.601	0.698

TABLE 1-3 Taxonomic Classification and Selected Chemical and Physical Properties of the Soils Named in Table 1-2

Soil[a]	Horizon	Taxonomic classification	pH	TOC (%)	Sum of cations CEC (cmolc/kg)	Exchange OH (cmolc/kg)	MnO2 (%)	Amorph. Fe2O3 (%)	Free Fe2O3 (%)	Al2O3 (%)	CaCO3 (%)	Sand (%)	Silt (%)	Clay (%)
Alligator	Ap	Very-fine, montmorillonitic, acid, thermic Vertic Haplaquept	4.8	1.54	30.2	3.5	0.028	0.33	0.74	0.15	—	5.9	39.4	54.7
Unnamed	Ap	Calciorthid	8.5	0.44	14.7	33.8	0.015	0.050	0.25	0.000	7.39	70.0	19.3	10.7
Cecil	Ap	Clayey, kaolinitic, thermic Typic Hapludult	5.7	0.61	2.0	0.011	0.099	1.76	0.27	—	67.7	12.8	7.3	
Cecil	B	Clayey, kaolinitic, thermic Typic Hapludult	5.4	0.26	2.4	6.6	0.002	0.082	7.48	0.94	—	30.0	18.8	51.2
Kula	Ap1	Medial, isothermic Typic Euthandept	5.9	6.62	22.5	82.4	0.093	1.68	5.85	3.51	—	73.7	25.4	0.9
Kula	Ap2	Medial, isothermic Typic Euthandept	6.2	6.98	27.0	58.5	0.13	1.64	6.95	3.67	—	66.6	32.9	0.5
Lafitte	Ap	Euic, thermic Typic Medisaprist	3.9	11.6	26.9	4.7	0.009	1.19	1.16	0.28	—	60.7	21.7	17.6
Molokai	Ap	Clayey, kaolinitic, isohyperthermic Typic Torrox	6.0	1.67	11.0	7.2	0.76	0.19	12.4	0.91	—	25.7	46.2	28.2
Norwood	Ap	Fine-silty, mixed (calc.), thermic Typic Udifluvent	6.9	0.21	4.1	0.0	0.008	0.061	0.30	0.016	—	79.2	18.1	2.8
Olivier	Ap	Fine-silty, mixed, thermic Aquic Fragiudalf	6.6	0.83	8.6	1.9	0.27	0.30	0.71	0.071	—	4.4	89.4	6.2
Unnamed	B21h	Spodosol	4.3	1.98	2.7	5.2	0.0000	0.009	0.008	0.22	—	90.2	6.0	3.8
Webster	Ap	Fine-loamy, mixed, mesic Typic Haplaquoll	7.6	4.39	48.1	14.1	0.063	0.19	0.55	0.10	3.14	27.5	48.6	23.9
Windsor	Ap	Mixed, mesic Typic Udipsamment	5.3	2.03	2.0	10.2	0.041	0.42	1.23	0.56	—	76.8	20.5	2.8
Windsor	B	Mixed, mesic	5.8	0.67	0.8	10.1	0.031	0.23	0.79	0.29	—	74.8	24.1	1.1

[a] The states from which the soil samples originated are Louisiana (Alligator, Lafitte, Norwood, and Olivier soils), South Carolina (Cecil soil), Hawaii (Kula and Molokai soils), Iowa (Webster soil), New Hampshire (Windsor soil), New Mexico (Calciorthid), and Florida (Spodosol).

retention, significant relationships among retention parameters and soil and element properties exist, even among soils with greatly different characteristics.

Langmuir

The Langmuir isotherm is the oldest and most commonly encountered in soils. It was developed to describe the adsorption of gases by solids where a finite number of adsorption sites in the surface is assumed (Langmuir, 1918). As a result, a major advantage of the Langmuir equation over linear and Freundlich types is that a maximum sorption capacity is incorporated into the formulation of the model, which may be regarded as a measure of the amount of available retention sites on the solid phase. The standard form of the Langmuir equation is

$$\frac{S}{S_{max}} = \frac{\omega C}{1 + \omega C} \tag{1-3}$$

where ω and S_{max} are adjustable parameters. Here ω (in mL g^{-1}) is a measure of the bond strength of molecules on the matrix surface and S_{max} (in μg g^{-1} of soil) is the maximum sorption capacity or total amount of available sites per unit soil mass. In an attempt to classify the various shapes of sorption isotherms, it was recognized that the Langmuir isotherm is the most commonly used and is referred to as the *L-curve isotherm* (Sposito, 1984).

The Langmuir sorption isotherm has been used extensively by scientists for several decades. Travis and Etnier (1981) provided a review of studies where the Langmuir isotherm was utilized to describe P retention for a wide range of soils. Moreover, Langmuir isotherms were used successfully to describe Cd, Cu, Pb, and Zn retention in soils. Figure 1-4 shows experimental and fitted isotherm examples of use of the Langmuir equation to describe Cu retention in Cecil and McLaren soil after 192 h of reaction.

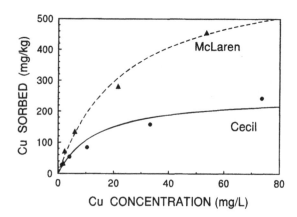

Figure 1-4   Retention isotherms for Cu after 8 days of reactions for Cecil and McLaren soils. Solid curves are calculated isotherms using the equilibrium Langmuir model.

Two-Site Langmuir

Based on several retention data sets, the presence of two types of surface sites responsible for the sorption of P in several soils was postulated. As a consequence, the Langmuir two-surface isotherm was proposed (Holford et al. 1974) such that:

$$\frac{S}{S_{max}} = \frac{F_1 \omega_1 C}{1 + \omega_1 C} + \frac{F_2 \omega_2 C}{1 + \omega_2 C} \tag{1-4}$$

where F_1 and F_2 (dimensionless) are considered as fractions of type 1 and type 2 sites to the total sites ($F_1 + F_2 = 1$), and ω_1 and ω_2 are the Langmuir coefficients associated with sites 1 and 2, respectively. The above equation is an adaptation of the original equation proposed by Holford et al. (1974) and was used to describe P isotherms by Holford and Mattingly (1975) for a wide range of soils.

A more recent adaptation of the two-surface Langmuir equation is the incorporation of a sigmoidicity term where

$$\frac{S}{S_{max}} = \frac{F_1 \omega_1 C}{1 + \omega_1 C + (\sigma_1/C)} + \frac{F_2 \omega_2 C}{1 + \omega_2 C + (\sigma_2/C)} \tag{1-5}$$

The terms σ_1 and σ_2 are the sigmoidicity coefficients (μg cm^{-3}) for type 1 and 2 sites, respectively. Schmidt and Sticher (1986) found that the introduction of this sigmoidicity term was desirable in order to adequately describe sorption isotherms at extremely low concentrations. Although the Langmuir approach has been used to model P retention and transport from renovated wastewater, the two-surface Langmuir with sigmoidicity approach has rarely been used to describe heavy metal retention during transport in soils.

GENERAL ISOTHERM EQUATIONS

Although heavy metal retention has been extensively studied for pure systems such as goethite, humic acids, and clay minerals, such a mechanism is not often easy to extend to soils (Kinniburgh et al., 1983; Kinniburgh, 1986). Soils are often viewed as complex mixture of substances, thus forming highly heterogeneous systems. Therefore, the description of heavy metal sorption in soils often remains empirical, such as the Freundlich and Langmuir models described above. As a result, "a general isotherm equation" describing the affinity of heavy metals to different binding sites on surfaces of soils was proposed:

$$\frac{S}{S_{max}} = \int_0^\infty w(\zeta) \, \Gamma(\zeta, C) \, d\zeta = f(C) \tag{1-6}$$

where $f(C)$ is a closed form isotherm expression as a function of concentration

C (mmol L^{-1}). The terms S and S_{max} denote the amount of solute sorbed and the maximum sorption capacity (mmol kg^{-1}), respectively. The function $w(\zeta)$ is equivalently defined as a weighting function (Sposito, 1984) or a site affinity distribution function (SADF) (Kinniburgh et al., 1983), and ξ is a an empirical affinity coefficient (dimensionless). The function w may also be viewed as a frequency distribution of the affinity coefficient ζ. This function w may be represented by a normal distribution. The function Γ is a local isotherm equation and is often represented by the Langmuir equation:

$$\Gamma(\xi, C) = \frac{\xi C}{1 + \xi C} \qquad (1\text{-}7)$$

A major advantage of the general isotherm formulation (eq. 1-6) is that it can be used to derive several commonly known isotherm reactions, including the Freundlich and Langmuir isotherms. As outlined by Kinniburgh et al. (1983) and later by Hinz et al. (1994), the following isotherms can be derived.

Case 1. By use of the Dirac delta function δ for $w(\zeta)$ as:

$$w(\zeta) = \delta(\zeta - k) \qquad (1\text{-}8)$$

and incorporation of eq. 1-8 into eq. 1-6, and integration yields the Langmuir isotherm:

$$\frac{S}{S_{max}} = \frac{kC}{1 + kC} \qquad (1\text{-}9)$$

where k is an "overall" affinity coefficient that is equivalent to ω of eq. 1-3.

Case 2. Selecting $w(\zeta)$ in the form:

$$w(\zeta) = F_1\delta(\zeta - k_1) + F_2\delta(\zeta - k_2) \qquad (1\text{-}10)$$

and proceeding as above yields the two-surface Langmuir isotherm equation:

$$\frac{S}{S_{max}} = \frac{F_1 k_1 C}{1 + k_1 C} + \frac{F_2 k_2 C}{1 + k_2 C} \qquad (1\text{-}11)$$

where F_1 and F_2 are the fraction of sites and $F_1 + F_2 = 1$. The numerical values of the "local" and "overall" affinity coefficients are identical for all Langmuir expressions.

Case 3. A hyperbolic weighting function for $w(\zeta)$ was proposed by Kinniburgh et al. (1983):

$$w(\zeta) = \frac{\zeta^\beta \sin(\pi\beta)}{\pi\zeta(\zeta - k)} \tag{1-12}$$

and yields the general Freundlich isotherm of the form

$$\frac{S}{S_{max}} = \left[\frac{kC}{1 + kC}\right]^\beta \tag{1-13}$$

where $0 < \beta < 1$ is a dimensionless heterogeneity factor that determines the spread of the distribution function $w(\zeta)$.

Case 4. In this case, the shape of the weighting function closely resembles the normal distribution (Sposito, 1981, 1984; Kinniburgh et al., 1983):

$$w(\zeta) = \frac{\sin(\pi\beta)}{\pi[k^{-\beta}\zeta^\beta + 2\cos(\pi\beta) + k^\beta\zeta^{-\beta}]} \tag{1-14}$$

This results in the general Langmuir–Freundlich isotherm expressed as:

$$\frac{S}{S_{max}} = \frac{(kC)^\beta}{1 + (kC)^\beta} \tag{1-15}$$

where $0 < \beta < 1$. The "overall" affinity coefficient represents the mean of the distribution function. A graphical distribution of the function $w(\zeta)$ for different values of β are shown in Figure 1-5.

Case 5. The Freundlich isotherm is derived from the general Langmuir-Freundlich eq. 1-15 for the limiting case when $1/k \gg C$:

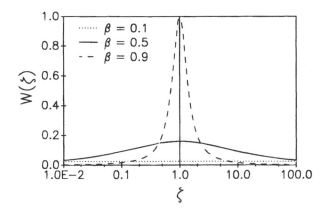

Figure 1-5 Weighting function of the general Freundlich (GF) isotherm for three different values of the heterogeneity factor β.

$$S = [S_{\max}k^\beta] \, C^\beta = k_f C^\beta \qquad (1\text{-}16)$$

where k_f and β are equivalent to K_f and b of eq. 1-1.

Several of the above isotherms have been applied to soils, soil minerals, and mixtures of ferrihydrite and humic acids for varying pHs (Kinniburgh et al., 1983; Kinniburgh, 1986; Altmann and Leckie, 1987). However, the Langmuir equation is perhaps the simplest form that is commonly used for describing solute sorption in soils and the Freundlich equation is most often used for heavy metal sorption. At low solution concentrations, the Freundlich equation, which can be derived from the general Freundlich–Langmuir equation, may be used. In practice, the use of the Freundlich equation is not restricted to low concentrations.

ASYMPTOTIC BEHAVIOR

Kinniburgh (1986) characterized different isotherms by their behavior at low and high concentrations. Specifically, two limiting criteria can be expressed. At low concentrations, we have

$$\lim_{C \to 0} \left[\frac{\partial S}{\partial C} \right] = \infty \qquad or \qquad \left. \frac{\partial S}{\partial S} \right|_{C=0} = constant \qquad (1\text{-}17)$$

At high C, we have

$$\lim_{C \to \infty} S(C) = \infty \qquad or \qquad \lim_{C \to \infty} S(C) = S_{\max} \qquad (1\text{-}18)$$

With the exception of the Freundlich isotherm, all cases of the general isotherm approach a constant value at high concentrations, usually denoted as the *sorption maximum*. Such sorption maxima may be related to the specific surface area of pure substances. At low C, the first derivative of the Langmuir and two-site Langmuir isotherms approach a constant, implying that sorption affinity approaches a constant value. In contrast, the first derivative of the general Freundlich, general Langmuir-Freundlich, and Freundlich equations approach infinity as C approaches zero. In other words, the affinity increases with decreasing C. Hinz and Selim (1994) demonstrated the asymptotic behavior for Zn and Cd isotherms for two soils. Examples for Zn isotherms for Windsor soil following 14 d adsorption are shown in Figures 1-6 and 1-7. Here, the results are presented using log scale. At low concentrations, Langmuir-type isotherms show a slope of unity on the log scale plot. Langmuir-type equations under-predicted the experimental results at low concentrations. Although there was no particular isotherm equation that successfully guaranteed described measured results, Hinz

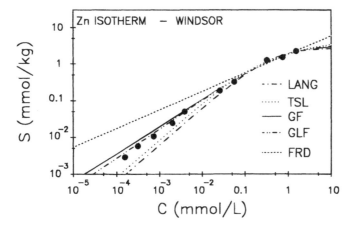

Figure 1-6   Isotherm data of Zn [concentration in soil solution (C) vs. the amount sorbed by the soil (S)] in Windsor soil after 14 days of reaction. Solid and dashed curves are fitted isotherm based on equations given in Table 1-1 (LANG, Langmuir; TSL, two-site Langmuir; GF, general Freundlich; GLF, general Langmuir-Freundlich; and FRD, Freundlich equations).

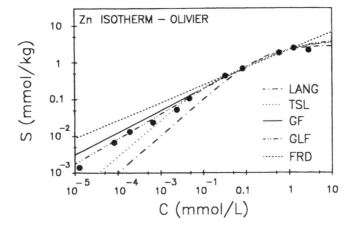

Figure 1-7   Isotherm data of Zn [concentration in soil solution (C) vs. the amount sorbed by the soil (S)] in Olivier soil after 14 days of reaction. Solid and dashed curves are fitted isotherm based on equations given in Table 1-1 (LANG, Langmuir; TSL, two-site Langmuir; GF, general Freundlich; GLF, general Langmuir-Freundlich; and FRD, Freundlich equations).

and Selim (1994) pointed out that a simple observation of the slope of the data allows an educated guess of the isotherm that may be best suitable for the experimental results.

Thermodynamics, through the use of state functions, describes the properties of systems at equilibrium. Kinetics, on the other hand, deals with descriptions of time-dependent processes, which are path-dependent. These two areas of science together constitute a powerful body of scientific law and theory that can be used to describe both systems and processes. A number of methods have been developed to obtain experimental kinetic data and analyze the data in order to arrive at a correct interpretation of the processes responsible for the observed time-dependent phenomena. This chapter reviews the various methods that have been developed for obtaining and analyzing kinetic data for soil systems. First, a few basic kinetic concepts must be introduced because they will be used in the discussions to follow.

BASIC KINETIC CONCEPTS

Consider the following sequence of reactions describing the surface complexation of an oxyanion by a metal oxide surface:

$$SOH(s) + H^+(aq) = SOH_2^+(s) \tag{2-1}$$

$$SOH_2^+(s) + L^{n-}(aq) = SOH_2^+ L^{n-}(s) \tag{2-2}$$

$$SOH_2^+ L^{n-}(s) = SL^{1-n}(s) + H_2O(l) \tag{2-3}$$

where SOH is the metal oxide surface hydroxyl group, SOH_2^+ is the protonated surface hydroxyl group, L is the oxyanion ligand of valence n, $SOH_2^+ L^{n-}$ is the outer-sphere metal–ligand complex, and SL^{1-n} is the inner-sphere metal–ligand complex. If these reactions occur as written at the molecular level, then they are **elementary** reactions. The series of elementary reactions shown above comprise the **mechanism** for the **overall** reaction:

$$SOH(s) + H^+(aq) + L^{n-}(aq) = SL^{1-n}(s) + H_2O(l) \qquad (2\text{-}4)$$

The rate of conversion (reaction rate) of each chemical species in the three elementary reaction steps can now be written as:

$$\frac{d[SOH]}{dt} = -k_1[SOH][H^+] + k_2[SOH_2^+] \qquad (2\text{-}5)$$

$$\frac{d[SOH_2^+]}{dt} = k_1[SOH][H^+] - k_2[SOH_2^+] - k_3[SOH_2^+][L^{n-}]$$

$$+ k_4[SOH_2^+L^{n-}] \quad (2\text{-}6)$$

$$\frac{d[L^{n-}]}{dt} = -k_3[SOH_2^+][L^{n-}] + k_4[SOH_2^+L^{n-}] \qquad (2\text{-}7)$$

$$\frac{d[SOH_2^+L^{n-}]}{dt} = k_3[SOH_2^+][L^{n-}] - k_4[SOH_2^+L^{n-}] - k_5[SOH_2^+L^{n-}]$$

$$+ k_6[SL^{1-n}][H_2O] \quad (2\text{-}8)$$

$$\frac{d[SL^{1-n}]}{dt} = k_5[SOH_2^+L^{n-}] - k_6[SL^{1-n}][H_2O] \qquad (2\text{-}9)$$

where the brackets refer to concentrations and the ks are the rate coefficients (h^{-1}).

Rate of conversion equations for H^+ and H_2O are not shown because they are trivial. These rate of conversion equations are known as **rate laws** but are not to be confused with fundamental laws of nature. They are best regarded as rate functions and will be referred to as such in subsequent discussions. Note that rate functions are written in terms of concentrations rather than thermodynamic activities. This is because spatial concentrations of colliding molecules determine molecular collision rates and hence reaction rates.

Assume that the first reaction (the protonation of the surface hydroxyl group) in the above mechanism occurs "instantaneously" so that this reaction is always at equilibrium. The rates of the forward and reverse reactions are equal at equilibrium; therefore,

$$k_1[SOH][H^+] = k_2[SOH_2^+] \qquad (2\text{-}10)$$

Rearranging this equation leads to:

$$K_1 = \frac{k_1}{k_2} = \frac{[SOH_2^+]}{[SOH][H^+]} \qquad (2\text{-}11)$$

where K_1 is the equilibrium constant for the reaction.

Thus, the fundamental link between thermodynamics and kinetics is established. The ratio of forward and reverse rate coefficients for an elementary reaction gives the equilibrium constant for that reaction. If the forward rate coefficient and the equilibrium constant for an elementary reaction are determined from kinetic and equilibrium experiments, then it is not necessary to measure the reverse rate coefficient. It can be calculated. However, this only applies to rate coefficients for elementary reactions, not for rate coefficients for kinetic processes that include both mass transfer and reaction kinetics.

Now assume that the last step in the reaction sequence (the formation of the inner-sphere metal–ligand complex) is irreversible and that the rate of formation of the outer-sphere metal–ligand complex is approximately equal to the rate of conversion of the outer-sphere complex to the inner-sphere complex $(d[SOH_2^+L^{n-}]/dt \simeq 0)$. Therefore,

$$k_3[SOH_2^+][L^{n-}] \simeq k_4[SOH_2^+L^{n-}] + k_5[SOH_2^+L^{n-}] \tag{2-12}$$

The concentration of the outer-sphere complex is therefore,

$$[SOH_2^+L^{n-}] = \frac{k_3}{k_4 + k_5}\,[SOH_2^+][L^{n-}] \tag{2-13}$$

Setting the rate of formation of the outer-sphere complex equal to its rate of conversion is known as the **steady-state approximation**, and the outer-sphere complex is a **reactive intermediate** under such conditions. A steady state occurs when only a single or some of the elementary reactions in a mechanism are at equilibrium. Complete equilibrium requires that the rates of forward and reverse reactions must be equal for all the elementary reactions and that all species must be at steady state. This is the **principle of detailed balancing** and is a consequence of the theory of **microscopic reversibility**, which requires that forward and reverse reactions in an elementary process follow the same path.

If the last step in the reaction sequence shown above is slow compared to the other two, then that reaction is the rate-determining step. If on the other hand the second reaction is slow compared to the others, then it is the rate-determining step. In a reaction sequence in which the reactions occur in series (consecutively), the slowest reaction is rate determining. Suppose however that an alternate reaction is possible, in which the inner-sphere complex is formed directly without going through the outer-sphere intermediate:

$$SOH_2^+(s) + L^{n-}(aq) = SL^{1-n}(s) + H_2O(l) \tag{2-14}$$

If this parallel (concurrent) reaction is faster than reaction (eq. 2-2) in the above reaction sequence, then it will be rate determining. If however, reaction (eq. 2-2) is faster, it will be rate determining. In a parallel reaction sequence, the fastest reaction determines the overall rate. Although slower concurrent reactions still occur, less reactants are consumed by these slower reactions than by the fastest parallel reaction.

It should be noted that the rate coefficients in eqs. 2-4 to 2-13 are conditional rate coefficients that vary with the pH and ionic strength of the solution in contact with the variable charge surface that forms the surface complex with the solute of interest. We will return to this point in Chapter 8.

There are many references (e.g., Frost and Pearson, 1961; Laidler, 1987; Gardiner, 1969; Bernasconi, 1986) that cover the basic concepts of chemical kinetics and the reader is encouraged to refer to some of these for more detailed and complete information. There is a good chapter on ion exchange kinetics in Helfferich's (1962) book on ion exchange. An excellent reference on kinetics of geochemical processes is available (Lasaga and Kirkpatrick, 1981). Until recently, reference books on kinetic processes in soils were not available. Sparks (1985, 1986, 1989) has produced two review chapters and a new book to fill this information gap. Sparks and Suarez (1981) edited a special publication on rates of chemical processes in soils. Sposito (1994) has revised and expanded upon material presented in his earlier texts (Sposito, 1981, 1984) by including a rigorous mathematical treatment of the kinetics of chemical reactions and diffusion in soils. With these concepts established, a brief discussion of the types of reactions that occur in soil systems is necessary because reaction type will often dictate the choice of method to be used.

TYPES OF REACTIONS IN SOIL ENVIRONMENTS

Soils and other geochemical systems are quite complex, and therefore there are different types of reactions that can occur concurrently and consecutively in these systems. Figure 2-1 illustrates several types of reactions that occur in soil systems and the time ranges (presented in readily identifiable time units) required to attain equilibrium by these reactions. The ion association, multivalent ion

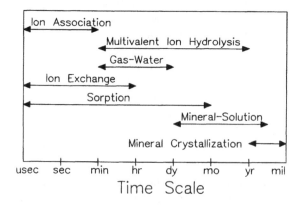

Figure 2-1   Time ranges required to attain equilibrium by different types of reactions in soil environments.

hydrolysis, and mineral crystallization reactions are all homogeneous because they occur within a single phase. The first two of these occur in the liquid phase, while the last occurs in the solid phase. The other reaction types are heterogeneous because they involve transfer of chemical species across interfaces between phases. Ion association reactions refer to ion pairing, complexation (inner- and outer-sphere), and chelation-type reactions in solution. Gas–water reactions refer to the exchange of gases across the air-liquid interface. Ion exchange reactions refer to electrostatic ion replacement reactions on charged solid surfaces. Sorption reactions refer to simple physical adsorption, surface complexation (inner- and outer-sphere), and surface precipitation reactions. Mineral–solution reactions refer to precipitation/dissolution reactions involving discrete mineral phases and coprecipitation reactions by which trace constituents can become incorporated into the structure of discrete mineral phases.

Reactions in soil environments encompass a wide range of time scales as Figure 2-1 shows. Furthermore, these reactions can occur concurrently and consecutively. The complexity of reactions in soils that occur over a time continuum defies a simple analysis of the kinetics involved (Sposito, 1986). Numerous methods have been developed to isolate and study the various types of reaction that occur in soils. The choice of method largely depends on the type of reaction to be studied, although some methods are applicable to more than one type of reaction.

The sequence of steps that may be followed in a typical series of kinetic studies on heterogeneous systems such as soils are as follows:

1. Select the kinetic method(s) to be used for the reaction to be studied (relaxation, batch, flow, stirred-flow methods).
2. Obtain the kinetic data under varying reactant concentrations, temperature, pH, ionic strength, and composition of other solution components.
3. Determine the rate function(s) from the experimental data using initial rate, isolation, graphical, rate coefficient constancy, fractional lives, or parameter optimization methods.
4. Propose mechanism(s) from experimental rate function and other data.
5. Test mechanism(s) by conducting experiments designed to eliminate alternative mechanisms.
6. Refine or reject mechanism(s).
7. Perform additional experiments as needed to validate or eliminate revised mechanism(s).

Rarely are all these steps followed in the course of a single study and published as a single paper. Many experiments are usually required to establish a mechanism, which is largely a trial-and-error process of testing and retesting to rule out alternate explanations for the observed rate functions. A lifetime of investigations may be required to establish a particularly complicated mechanism or it may not yield to solution even then. In the following section, some of the various methods for obtaining kinetic data will be described. The advantages and limitations of each method and its overall suitability for studying the various types of reactions are also discussed in some detail.

METHODS OF OBTAINING KINETIC DATA

Experimental methods for obtaining kinetic data for studying various types of reactions in soils fall into three broad categories: (1) relaxation methods, (2) batch methods, and (3) flow methods. Flow methods can be further subdivided into flow methods without mixing and flow methods with mixing. A review of these methods is found in Sparks (1989).

Relaxation Methods

Relaxation methods were developed for studying fast reactions (those occurring on time scales of milliseconds, microseconds, or less) such as ion association reactions in solution and the more rapid ion exchange and surface complexation reactions on solid surfaces. They involve the application of a perturbation (e.g., pressure, temperature, concentration jumps; electric field pulses) to a system at equilibrium and then following the return to equilibrium by monitoring some property of the system (e.g., conductivity or fluorescence). The application of relaxation methods to the study of reactions on soil constituents has been presented by Sparks and Zhang (1991) and will not be covered here. This chapter will focus on batch and flow methods that are suitable for studying slower reactions in soils (those occurring on time scales of seconds to years or more). Sorption reactions on soil surfaces may occur over long periods of time. Although the kinetics of the most rapid of these reactions can only be studied by relaxation methods, batch and flow methods are applicable to the study of reactions in which loss of solute from solution onto soil surfaces occurs over longer time periods. Experimental data for the applications of the kinetic models presented in subsequent chapters were obtained using some of these methods.

Batch Methods

Batch methods have long been used to obtain kinetic data for reactions in soils. A unique feature of batch methods is that they are primarily a closed system. After all the reactants are added and mixed together, no additional amounts of reactants are added and products are allowed to accumulate. The only material removed from the system is that removed for analysis, and that is usually a very small fraction of the total. Thus, sampling perturbations are typically minor. A typical kinetic experiment using a batch reactor may be performed as follows:

1. Suspend a known amount of the solid phase to be studied in a known volume of background electrolyte solution or other solution of known composition depending

on requirements. A hydration period of several hours may be required for the solid phase before beginning the experiment.

2. Begin mixing and flow of any gases used to purge the batch reactor and adjust the pH to the desired level.
3. At some initial or starting time ($t = 0$), add a known amount of solute of interest to achieve desired initial reactant concentrations, ionic strength, and total volume.
4. At periodic intervals, withdraw portions of the suspension using a syringe sampler and filter the suspension samples through membrane filters.
5. Analyze the filtrates and separated solid phase as needed.

Numerous variations are of course possible, but the steps outlined in the flow diagram are fairly typical. Reactor design, mixing techniques, and techniques for separating the solid and liquid phases are all critical to obtaining good data and each will be discussed in turn. Special techniques include *in situ* monitoring of reactant activities with ion-selective electrodes, which renders phase separation unnecessary, and the use of radionuclides to monitor the extent of the reaction.

Batch Reactor Design

A typical batch reactor configuration is shown in Figure 2-2. The reactor itself is a cylinder constructed of glass or plastic. The best material for minimizing solute sorption onto the reactor walls should be selected. The choice of material is normally based on the solute to be studied. Glass is usually more suited for use with organic solutes, and plastic is usually more suited for some inorganic species (e.g., trace metals). A glass or plastic Erlenmeyer flask makes a suitable low-cost reactor.

The size of the reactor is largely dictated by the volume and number of samples to be taken. The reactor should be of sufficient size to obtain a sufficient number of samples to adequately describe the kinetics of the reaction and the volume of each sample must be sufficient to obtain accurate analytical measurements. Normally, a one- to two-liter reactor volume is adequate for most studies. The reactor contents can be mixed with an overhead stirrer or from below with a magnetic stirrer. Mixing techniques are critical and are discussed in more detail below.

If accurate temperature control is required, the reactor may be surrounded with a jacket through which a fluid (normally water) is pumped to maintain a constant temperature. Alternatively, the reactor can be placed in a constant-temperature bath. When temperature must be monitored, the reactor must be equipped with a thermometer as temperatures inside the reactor may vary slightly from the constant temperature bath or jacket because of reactions and mixing.

Batch reactors are often equipped with gas dispersion tubes through which an inert gas such as nitrogen or argon is bubbled through the suspension. This is only necessary if one wants to sweep CO_2 and O_2 out of the suspension. If complete exclusion of CO_2 and O_2 is required or if exact control of the gas composition in the reactor headspace is needed, the reactor must be completely

Batch Reactor

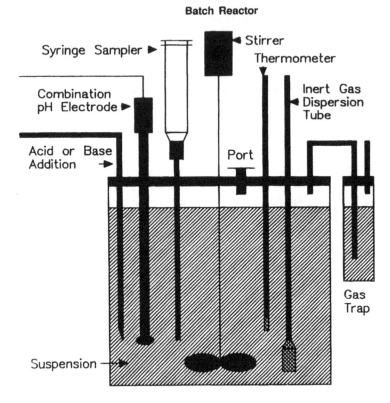

Figure 2-2 Typical batch reactor configuration. pH is controlled by a combination pH electrode and automatic burette connected to an autotitrator, a syringe sampler allows for removal of a subsample of suspension, an addition port permits injection of solute, an inert gas is bubbled through the suspension by means of a gas dispersion tube, and the system is vented through a gas trap; a thermometer allows for temperature monitoring, and the suspension is mixed with an overhead stirrer.

sealed from the atmosphere and a gas trap must be used to vent the purge gas and to prevent contact between the reactor headspace and the outside atmosphere. This is normally required only when very precise control over redox status is needed.

An addition port is provided for the addition of the soluble reactants or other solutes. A simple funnel may be used if the suspension is open to the atmosphere; otherwise, a serum stopper for use with a syringe needle is required. The syringe sampler consists of a tube into the suspension, the top of which is constructed to allow connection to a Luer-lock syringe. Syringe size depends on the volume to be sampled, but normally a 10-mL plastic syringe is used. The rubber end of the syringe plunger can sometimes cause contamination problems, especially with Zn. More costly Teflon plunger syringes are needed in such cases. The syringe should be filled with inert gas prior to sampling in order to avoid introducing CO_2 or O_2 into the suspension during sampling if these gases must be excluded.

A pH electrode connected to an autotitration instrument (pH meter, controller, and autoburette) is often used to maintain a constant pH (pH-stat) during the reaction. If it is desired to monitor pH changes during the reaction, then the pH meter can be connected to a recorder. It should be remembered that autotitration units have finite response times that are longer than many surface-controlled reactions. Thus, pH changes can occur before the autotitration system can fully respond, and there will be some delay between the reaction and return to pH-stat conditions.

A batch reactor of the type just described was first developed and used by Patrick et al. (1973) to study soils under controlled pH and redox conditions. Although not developed to collect kinetic data per se, the Patrick et al. (1973) batch reactor was the forerunner of subsequently developed kinetic batch reactors. Zasoski and Burau (1978) developed a batch reactor designed specifically for sorption kinetic experiments. The use of the Zasoski and Burau (1978) reactor to obtain kinetic data largely follows the sequence outlined in Figure 2-3. Harter and Lehmann (1983) used a Zasoski and Burau (1978) type reactor to separate rapid ion exchange (or surface complexation) reactions involving Ni and Cu from slower retention reactions. van Riemsdijk and Lyklema (1980a,b) and van Riemsdijk and de Hann (1981) used a batch reactor under pH-stat and P-stat conditions to study P retention kinetics by soils and soil constituents under constant supersaturation with respect to metal phosphates. Similarly, Phelan and Mattigod (1987) used a pH-stat and Ca-stat reactor to study the kinetics of P precipitation from supersaturated solutions. These are good examples of special-

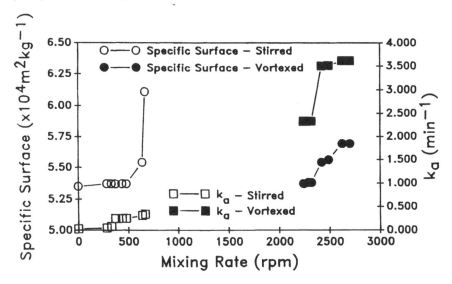

Figure 2-3 Effect of type and rate of mixing on soil surface area and apparent rate coefficients for K exchange. Specific surface is indicated by circles and rate coefficients by squares. Open symbols signify the stirred system, whereas filled symbols signify the vortex mixed system. (Adapted from Ogwada, 1986, *Soil Sci. Soc. Am. J.* 50:1158-1162.)

ized adaptations of the basic batch reactor to study precipitation kinetics. Amacher and Baker (1982) used a Zasoski and Burau (1978) type reactor to study the kinetics of Cr(III) oxidation by soils and MnO_2 minerals and the kinetics of Cr(VI) reduction by fulvic acid.

Mixing Techniques

The method of mixing and the mixing rate in a batch reactor are critical to obtaining consistent results. The rates of many, if not most, and perhaps all surface-controlled reactions such as ion exchange and surface complexation may be controlled by mass transfer processes such as diffusion through the hydrodynamic film surrounding the solid particles (film diffusion), by diffusion into or through the particle (particle diffusion), or by a combination of the two diffusion processes. Diffusion control occurs when the rate of transfer of the solute to the particle surface is rate limiting (i.e., slower than the more rapid ion exchange or surface complexation reaction).

Diffusion-controlled ion exchange was first demonstrated for ion exchange reactions of univalent cations on synthetic ion exchange resins (Boyd et al., 1947). Boyd et al. (1947) also developed the mathematical equations to describe diffusion-controlled ion exchange kinetics. Numerous criteria have been used to distinguish between film and particle diffusion-controlled kinetics including mixing rate, form of the kinetic equation obeyed by the data, and temperature dependence of the kinetics; but perhaps the most reliable method has been the use of the **interruption test** (Kressman and Kitchener, 1949). In this test, the ion exchange material is enclosed in a wire mesh cage attached to the end of a centrifugal stirrer. The reaction is begun by lowering the rotating stirrer with the ion exchange material in the cage into the solution containing the solute of interest. After a timed interval, the reaction is "interrupted" by raising the still rotating stirrer out of the solution. Analysis of the ion exchange material and solution phases gives the extent of the exchange reaction for the time interval. The process is repeated at various time intervals. If particle diffusion is rate limiting, the ion exchange rate immediately after reimmersion of the particles is greater than before interruption. If film diffusion is rate limiting, then interruption has no effect on reaction rate.

The wire cage apparatus is usable if the particle size of the exchange material is large enough to be contained in the cage whose openings must be large enough so that the solution will flow rapidly through the cage. Thus, the wire cage apparatus cannot be used for clay-sized particles. To solve this problem, Bunzl (1974) attached peat particles to a polyvinylchloride cylinder with a small propeller at the lower end. They were then able to use the interruption test to show that the rate of Pb-H exchange on peat is film diffusion controlled.

In a batch reactor of the type described above, it is desirable to reduce or eliminate film diffusion by vigorous mixing of the suspension. If mixing is sufficiently vigorous, particle diffusion may be reduced or eliminated as well.

Kressman and Kitchener (1949) showed that mixing was optimum at a stirring speed of 1000 to 1100 rpm for ion exchange in the synthetic resin system.

Ogwada and Sparks (1986c) investigated the effect of type and rate of agitation on the kinetics of K sorption by a Chester loam. Their original paper presents the data in tabular form. To clearly show the effect of type and rate of agitation on the specific surface of the soil and their measured apparent rate coefficient for sorption, their results are graphed in Figure 2-3.

In the stirred system k_a, the apparent rate coefficient for K sorption, increased steadily with increasing stirring rate. A sharp increase in specific surface was observed at the highest stirring speeds, indicating that some particle abrasion was occurring. The increase in surface area could account for an increase in reaction rate. However, k_a also increased when specific surface did not, indicating that the extent of film diffusion control on reaction rate diminished with increasing stirring rate.

The type of agitation had a very large effect on k_a, but not on specific surface. In the vortex mixed system, specific surface did increase with vortex rate, but the increase was not as large as in the stirred system. Furthermore, the specific surface in the vortex system at the lower vortex rates was the same as that in the stirred system at the lower stirring rates. Prolonged agitation periods are probably required to produce measurable effects on surface area and reaction rate. The difference in k_a values between the stirred and vortex mixed systems and the dependence of k_a on mixing rate in each mixing system indicates that film diffusion was probably controlling reaction rate in the stirred system, that film diffusion control was diminishing as stirring rate increased, that film diffusion was eliminated in the vortex system and now particle diffusion controlled reaction rate, and that particle diffusion control diminished with increasing vortex mixing rate.

Studies of the type conducted by Ogwada and Sparks (1986c) are rare and this is unfortunate given the demonstrated extreme importance of the type of mixing and the mixing rate on the kinetic results. More quantitative data of the type produced by Ogwada and Sparks (1986c) is needed. Although the effect of agitation rate on the kinetics of P retention in soils has been documented (Barrow and Shaw, 1979), similar data for other solutes and for various soil constituents are lacking. It is expected that the mineralogy of the soil will play an important role in determining whether or not the type and rate of agitation will measurably affect surface area and reaction rates.

Ogwada and Sparks (1986c) used the differences in observed rate coefficients between static (no mixing), stirred, and vortex mixed systems to develop a method for separating and quantifying the rate-limiting step in ion exchange reactions on soils and clays. By assuming that the steps in the ion exchange process occurred in series (film diffusion $\rightarrow$ particle diffusion $\rightarrow$ exchange reaction) and by using the steady-state approximation (mass transfer rate $\simeq$ exchange rate), they developed additive resistance relationships using reciprocals of the rate coefficients for each of the mixing systems:

$$\text{Static: } \frac{1}{k_{OS}} \simeq \frac{1}{k_R} + \frac{1}{k_P} + \frac{1}{k_F} \tag{2-15}$$

$$\text{Stirred: } \frac{1}{k_{OT}} \simeq \frac{1}{k_R} + \frac{1}{k_P} \tag{2-16}$$

$$\text{Vortex: } \frac{1}{k_{OV}} \simeq \frac{1}{k_R} \tag{2-17}$$

where k_{OS} is the observed rate coefficient in the static system with no mixing, k_{OT} is the observed rate coefficient in the stirred system, k_{OV} is the observed rate coefficient in the vortex mixed system, k_R is the rate coefficient for the exchange reaction, k_P is the rate coefficient for particle diffusion, and k_F is the rate coefficient for film diffusion.

First-order kinetics is assumed throughout and the reciprocals of the k values are the additive resistances for the mass transfer and exchange processes. Since k_{OS}, k_{OT}, and k_{OV} are determined experimentally, the unknown k_F and k_P values can be calculated by suitable rearrangement of the additive resistance equations. Ogwada and Sparks (1986b) used this approach to show that the rate of K exchange on kaolinite was film diffusion controlled under static conditions, but the rate of K exchange on vermiculite was both film and particle diffusion controlled under static conditions.

Film diffusion and particle diffusion are affected by numerous experimental conditions (Sparks, 1989). In batch systems, mixing greatly influences diffusion as already seen with vigorous mixing tending to reduce or eliminate diffusion control. In flow systems to be discussed in a later section, flow velocity affects film and particle diffusion. Hydrodynamic film thickness also affects film and particle diffusion. A decrease in film thickness favors particle diffusion-controlled kinetics. Hydrodynamic film thickness is in turn affected by the type and rate of mixing in a batch reactor, flow velocity in a flow system, hydration of cations, and ionic strength of the background electrolyte solution. Particle size also affects film and particle diffusion. Film diffusion usually predominates with small particles. Ion concentration is also important. Film diffusion is usually favored in solutions of low ionic concentration.

The previous discussion has established the importance of mixing in batch reactors. Mixing in batch reactors of the type described above is normally accomplished in one of two ways: from below using a stir bar driven by a magnetic stirrer, or from above using a propeller connected to a stirring motor. Either method can promote effective and vigorous agitation that will reduce or eliminate film diffusion. Reduction or elimination of particle diffusion, however, may not be possible in stirred reactors as the mixing rate is usually not as vigorous as with a vortex mixer.

The stir bar and magnetic stirrer approach is probably the simplest and easiest to use. Heating of the reactor from the stir plate can be eliminated by placing a heat shield between the stir plate and reactor and by enclosing the reactor in a constant-temperature jacket. Mixing with a stir bar over long periods of time in a glass reactor suffers from the disadvantage of having quartz introduced into the reactor by the grinding action of the stir bar on the glass. The overhead stirrer will eliminate this problem, but propeller design is critical to obtaining good mixing action. The propeller must be positioned near the bottom of the reactor to keep the particles in suspension. Mixing must be vigorous enough to keep sand-sized particles in suspension and a vortexing action must be created to pull organic matter under the surface that otherwise would float on top. The stratification of soil constituents in batch reactors because of weight differences is a recurring problem and satisfactory results cannot always be obtained with some soils. In a batch reactor that must be sealed against the outside atmosphere, mixing is most easily accomplished from below with a magnetic stirrer. However, specially designed propellers with bearings and gaskets are available to stir sealed reactors from above. Regardless of the method used, mixing should be as complete, uniform, and vigorous as possible with no unmixed zones in the reactor. When the solute reactant is added, mixing should be complete in less than 5 sec. Dyes or other colored solutions can be used to test the mixing process by visual observation.

Many batch experiments are of course not conducted in reactors as elaborate as that described above. When centrifuge tubes or Erlenmeyer flasks are used as simple batch reactors, mixing is normally accomplished by vortexing, reciprocal shaking, or end-over-end shaking in the case of tubes or by rotational or wrist-action shaking in the case of flasks.

Separation of Solid and Liquid Phases

This aspect of batch reaction methods is also critical to obtaining good results. Rapid filtration is normally used to separate the solid and liquid phases when samples are removed from a typical batch reactor. When rapid reactions are studied, sampling with the syringe sampler must be rapid and reproducible since the reaction is continuing as the sample is taken. A 5-sec sampling time is about optimal, but it should not exceed 10-sec, including connection of the syringe to the filter holder. Similarly, filtration time is also critical since the reaction is still proceeding as filtration occurs. Reproducible filtration times are mandatory for obtaining consistent results. The filtration time will depend on the nature of the solid phase as well as the solution:solid ratio. Well-dispersed solid phases with small particle sizes can take longer to filter. Short filtration times are favored by wide solution:soil ratios. One should strive for the lowest, most consistently reproducible filtration time. The pore size of the filter also affects filtration time, with smaller pores yielding longer filtration times. Normally, 0.2 or 0.45-μm membrane filters are used. The filters and holders must be chemically inert and should not sorb the solute of interest. They also should not release contaminants into the filtrate. Polycarbonate filters have been found to produce very consistent results in this regard.

The volume of the sampled solution and the weight of the separated solid phase can be determined as a check on whether the original solution:solid ratio in the batch reactor is maintained in the sample. Also, the separated solid phase can be analyzed as a check on the mass balance of the solute in the system. Zasoski and Burau (1978) demonstrated that the solution:solid ratio was not significantly altered by repeated sampling in a batch reactor.

For the study of slower reactions, centrifugation is acceptable for separating the solid and liquid phases. However, the reaction will still continue across the solid:liquid interface, although at a reduced rate. Centrifugation is valid if the extent of the reaction changes little during the time required for centrifuging and sampling.

In Situ *Methods*

In some cases, separation of solid and liquid phases can be avoided by using *in situ* analysis techniques. The only readily available method for monitoring solutes in batch reactors is the use of ion-selective electrodes. Spectroscopic methods for studying the kinetics of solute reactions on solid surfaces are still largely in the developmental stage. The subject of *in situ* analysis has already been introduced by discussion of the pH-stat characteristics of the batch reactor. Reactions at electrode surfaces are much the same as reactions at the surfaces of soil constituents. Electrodes have finite response times that may exceed or are of the same order of magnitude as rapid reactions at soil surfaces. Therefore, the reaction at the soil surface can be largely complete before the electrode can respond to changes in solute activity. Despite this limitation, *in situ* methods with ion-selective electrodes are to be preferred over phase separation methods where appropriate, because extra steps with their own sources of experimental error are eliminated. A serious limitation to the use of some ion-selective electrodes with soils is the presence of interferences in the soil solution. Interferences are much more readily controlled in studies with pure soil constituents where the chemical composition is known and can be controlled. Ion-selective electrodes with the greatest potential for *in situ* kinetic studies include those for NH_4^+, K^+, Ca^{2+}, Cu^{2+}, Cd^{2+}, and Pb^{2+}. Ogwada and Sparks (1986b) used a K^+ electrode to study the kinetics of K ion exchange. Aringhieri et al. (1985) used Cd^{2+} and Cu^{2+} electrodes to study the kinetics of Cd and Cu retention by a histosol. Jopony and Young (1987) used a Cu^{2+} electrode to study Cu desorption kinetics in the presence of an EDTA sink that lowered Cu^{2+} activity and thus initiated desorption.

Radiotracer Methods

This is a specialized technique and thus is considered separately. Radiotracer methods are an excellent means of following the extent of kinetic reactions. The radiotracer is a radionuclide of the solute of interest. Addition of the radiotracer to the system results in isotopic dilution of the chemical species of interest. It is assumed that the radiotracer behaves chemically in an identical fashion to the stable radionuclide. Because radiotracer analysis can be sensitive, extremely low concentrations of solute can be monitored with relative ease. Radiotracer concentrations can be chosen to yield optimal ratios of radiotracer activity to background activity. Corrections for radioactive decay are avoided by counting standards and samples at the same time. Another major advantage is that small aliquots (<1 mL) of solution can be taken for analysis. This makes the method particularly suited for studying reactions in small batch reactors such as centrifuge tubes. Small aliquots of solution removed for analysis alter the solution:soil ratio only slightly, and amounts of solute sorbed by the solid phase are readily corrected for the very small amount removed by sampling.

Aqueous concentrations of the solute of interest are calculated from the radionuclide activities of the samples and initial solutions (standards):

$$C = C_o \left[\frac{A - bkg}{A_o - bkg} \right] \tag{2-18}$$

where C is the solute concentration in the aqueous solution in contact with the soil, C_o is the initial solute concentration before reaction with the soil, A is the radiotracer activity in the aqueous solution in contact with the soil, A_o is the radionuclide activity in the initial solution before reaction with the soil, and bkg is the background radioactivity.

The concentration of solute in the soil phase at each sampling time is calculated from the difference between the initial aqueous solute concentration and the aqueous solute concentration at each sampling time:

$$S = (C_o - C) \frac{V}{W} \tag{2-19}$$

where S is the solute concentration in the soil (mol kg^{-1} or mg kg^{-1}), C_o is the initial aqueous solute concentration (mol L^{-1} or mg L^{-1}), C is the aqueous solute concentration at time t, V is the volume of solution (L), and W is the weight of the soil (kg).

A correction is made for removal of a small amount of sample for counting (Amacher et al., 1986).

Radiotracers have long been used to study ion exchange reactions at or near equilibrium. An example of their use in kinetic studies was provided by Amacher et al. (1986) who studied the kinetics of Cr(VI), Cd, and HgCl$_2$ retention by soils using radiotracers to follow the extent of the reactions. A flow diagram of the batch method used by Amacher et al. (1986) to obtain the kinetic retention data used in the model applications presented in subsequent chapters is shown in Figure 2-4. This method is a relatively simple approach for studying slower retention reactions in that the reactions were carried out in centrifuge tubes, soil and aqueous phases were separated by centrifugation, and radiotracers were used to follow the reactions.

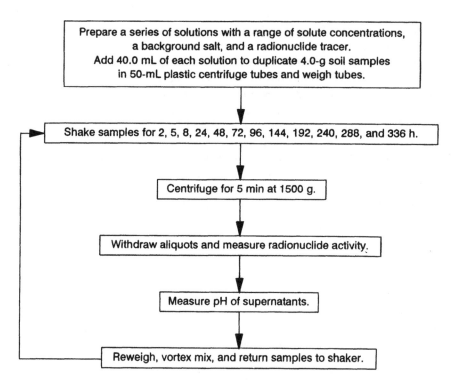

Figure 2-4 Flow chart for batch retention studies in centrifuge tubes using radionuclide tracers.

Advantages and Limitations

Batch methods for obtaining kinetic data have a number of advantages, including:

1. There are generally low-cost equipment requirements and batch reactors are relatively easy to use.
2. Film diffusion and sometimes particle diffusion is eliminated by sufficiently vigorous mixing.
3. Constant solution:solid ratio is readily maintained in some batch reactor systems.
4. Reaction conditions such as pH, ionic strength, and removal of CO_2 and O_2 are easily controlled.

Batch methods also have a number of limitations, including:

1. Desorbed species are not removed and are allowed to accumulate in the inherently closed system of the batch reactor. Thus, unless a unidirectional reaction is being studied, reverse reactions must be taken into account in the data analysis. The accumulation of desorbed species can sometimes result in secondary precipitation reactions, which further complicate data analysis.
2. The mixing method employed may not produce a uniform suspension and may

not be sufficient to limit mass transfer of solute. Furthermore, surface area may be increased by prolonged mixing by some methods.

3. Sampling and phase separation steps are not always uniform and are operator dependent. In the case of rapid surface reactions, sampling and phase separation are not rapid enough to follow the reaction.

Despite these limitations, batch methods will probably continue to be used to study kinetic reactions in soils where appropriate, owing largely to their simplicity.

One of the simplest of all batch reactors is the centrifuge tube, and many kinetic and equilibrium solute sorption studies with soils and soil components have been conducted in centrifuge tubes. A descriptive diagram showing a typical sorption/desorption study carried out using centrifuge tubes is shown in Figure 2-5. Centrifuge tubes can be used effectively to obtain kinetic data if the following criteria are met:

1. The reaction(s) is sufficiently slow such that the reactant concentrations will not change appreciably during centrifugation and sampling steps. Even in the case of rapid reactions, this traditional batch method can still be used to study sorption after attainment of steady-state conditions.

2. Very small aliquots of sample are taken so that the solution:soil ratio is only slightly altered. Microanalytical or radiotracer techniques permit this approach.

3. pH does not need to be continuously monitored and adjusted. Adjustments in pH or pH measurements are possible with this batch method, but they can only be done at discrete time intervals. Often this is sufficient.

4. Continuous purging with inert gas is not required. Samples can be purged initially and at subsequent intervals if necessary. Tubes can be opened in a controlled atmosphere glove box to avoid contact with CO_2 or O_2 if required.

Batch

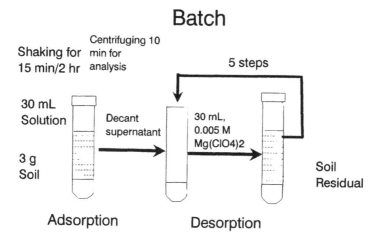

Figure 2-5 Descriptive diagram of batch sorption–desorption studies in centrifuge tubes.

Flow Methods

To overcome some of the problems associated with batch methods, flow methods have been developed and used to obtain data on kinetic reactions in soils and soil constituents. A review of these methods is given by Sparks (1989). A summary of some of the kinetic studies using flow methods is given in Table 2-1.

A unique feature of flow methods, as contrasted with batch methods, is that flow methods are inherently open systems. Solute is continually added to the system with flow methods and reaction products are continually removed. This unique feature produces special attributes for the flow methods that solve some of the problems associated with batch systems, but also involve some limitations as well. Perhaps the chief advantage of flow methods is the continual removal of reaction products that in batch reactors are allowed to accumulate. Another unique attribute of flow methods is that the solution:soil ratio is much narrower than that used in batch reactors. In the Zasoski and Burau (1978) type batch reactor, solution:soil ratios are typically of the order of 100:1, whereas in flow type reactors the ratio is typically less than 1:1. Narrow solution:soil ratios are sometimes used in simple centrifuge tube batch reactors where ratios from 1:1

TABLE 2-1 Summary of Studies Using Flow Methods to Obtain Kinetic and Sorption Isotherm Data

Flow method	Reaction studied	Reference
Miscible displacement with thin disk	Al/NH₄ and K/NH₄ exchange on soils	Sivasubramaniam and Talibudeen (1972)
	K exchange on soils and clay minerals	Sparks et al. (1980)
		Sparks and Jardine (1981)
		Sparks and Rechcigl (1982)
		Jardine and Sparks (1984)
		Sparks and Jardine (1984)
		Ogwada and Sparks (1986a,b,c)
	Al sorption by clay minerals and peat	Jardine et al. (1985a,b)
	Data analysis theory and methods	Skopp and McCallister (1986)
	SO₄ sorption-desorption in soils	Hodges and Johnson (1987)
	P and Si sorption on goethite	Miller et al. (1989)
Stirred-flow reactor	K exchange on kaolinite	Carski and Sparks (1985)
	NH₄ release from soils	Carski and Sparks (1987)
	Zn, Cd, and Hg sorption by humic acids	Randle and Hartmann (1987)
	Data analysis theory and methods	Schnabel and Fitting (1988)
	Ca/Mg exchange on soils	Seyfried et al. (1989)
	NO₃ sorption-desorption on soils	Toner et al. (1989)
Fluidized bed	Albite weathering	Chou and Wollast (1974)
Soil column	P sorption by soils under P-stat conditions	van Riemsdijk and van der Linden (1984)

to 10:1 are often used. Furthermore, in flow reactors, the solid phase will be reacted with a greater mass of solute (concentration $\times$ flow velocity $\times$ time) than in batch reactors (concentration $\times$ volume).

Two basic types of flow methods can be distinguished: those with mixing and those without. The chief limitation of unmixed flow reactors is that mass transfer processes are frequently limiting and, in the case of fast chemical reactions, are probably always limiting. Stirred-flow reactors and fluidized bed reactors may often overcome the mass transfer limitation and indeed these hybrid techniques may represent the best attributes of batch and flow methods. Each of these approaches is considered in turn.

Thin Disk Method

This method is also referred to as the miscible displacement or continuous flow method. In this method, a thin disk of dispersed solid phase is deposited on a porous membrane and placed in a holder. A pump is used to maintain a constant flow velocity of solution through the thin disk and a fraction collector is used to collect effluent aliquots. A thin disk is used in an attempt to minimize diffusion resistances in the solid phase. Disk thickness, disk hydraulic conductivity, and membrane permeability determine the range of flow velocities that are achievable. Dispersion of the solid phase is necessary so that the transit time for a solute molecule is the same at all points in the disk. However, the presence of varying particle sizes and hence pore sizes may produce nonuniform solute transit times (Skopp and McCallister, 1986). This is more likely to occur with whole soils than with clay-sized particles of soil constituents. Typically, 1- or 2-gram samples are used in kinetic studies on soils with the thin disk method, but disk thicknesses have not been measured. In their study of the kinetics of phosphate and silicate retention by goethite, Miller et al. (1989) estimated the thickness of the goethite disk to be 80 μm.

The available evidence suggests that thin disks do not minimize mass transfer processes. Comparison of the thin disk flow method with batch methods have shown that apparent rate coefficients are larger in batch systems and that steady-state conditions are obtained sooner (Sparks and Rechcigl, 1982; Ogwada and Sparks, 1986a). In addition, apparent rate coefficients obtained from the thin disk method may depend on the flow velocity through the disk. Thus, thin disk methods give rate coefficients for overall kinetic processes including mass transfer and reaction steps.

A problem with the thin disk method uncovered by Carski and Sparks (1985) is that the influent solution can be diluted by the solution used to load the solid phase onto the membrane filter, or the washing out of leftover sorbing solution during desorption can produce concentration changes not due to sorption or desorption. The dilution and washout effects could erroneously lead one to conclude that sorption and desorption has occurred when in fact there has been no reaction with the solid phase at all. This dilution and washout effect was demonstrated by Carski and Sparks (1985) for boron flowing through acid-washed sand where no retention was expected to occur.

An additional problem with the thin disk method is that control over reaction conditions within the thin disk is not usually possible. Although pH and other solution composition variables can be controlled in the influent solution, once the solution contacts the thin disk, direct control with feedback is no longer possible.

Stirred-Flow Method

Stirred-flow methods have long been used by chemical engineers and chemists to obtain kinetic data in homogeneous systems. They have only comparatively recently been used to obtain kinetic data in heterogeneous soil systems (Carski and Sparks, 1985; Randle and Hartmann, 1987; Seyfried et al., 1989; Miller et al., 1988). This method is often referred to as a continuously stirred tank reactor (CSTR) in chemical engineering and geochemistry literature.

The experimental setup for this method is shown in Figure 2-6. It is identical to the setup for the thin disk method except that the stirred-flow reactor is used

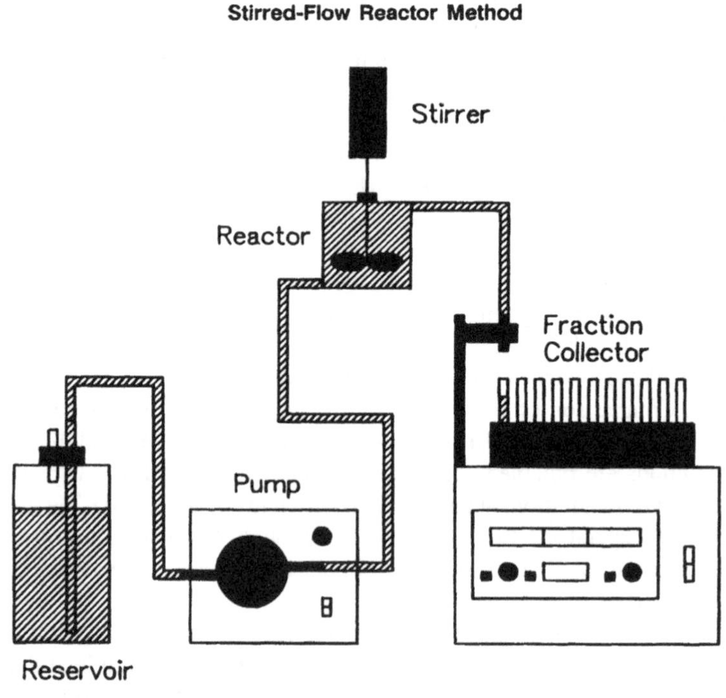

Stirred-Flow Reactor Method

Figure 2-6 Stirred-flow reactor method experimental setup. Background solution and solute are pumped from the reservoir through the stirred reactor containing the solid phase and are collected as aliquots by the fraction collector. Separation of solid and aqueous phases is accomplished by a membrane filter at the outlet end of the stirred reactor.

in place of the thin disk. Flow through the reactor is maintained at a constant velocity by a pump and a fraction collector is again used to collect reactor effluent.

Various designs of the stirred-flow reactor are possible. Carski and Sparks (1985) developed a relatively simple stirred-flow reactor constructed from a plastic syringe and membrane filter holder. The volume of the reactor is adjustable to allow one to add and maintain a known amount of solution to a known amount of solid phase. Mixing is accomplished by a magnetic stirrer.

Miller et al. (1988) developed a stirred-flow reactor in which stirring is accomplished by a propeller connected to a high-torque motor. A special bearing allows the propeller shaft to enter the reactor, but seals the reactor against leaks during solution flow.

Stirred-flow reactors retain all the advantages of flow methods in general and eliminate all the problems associated with the thin disk method. They also retain many advantages of the batch method. Reaction products desorbed into solution are continually removed. Film or particle diffusion is reduced or eliminated by mixing within the reactor. Direct control with feedback of reactor conditions is possible. Although stirred-flow reactors normally have much smaller volumes than conventional batch reactors, it is possible to construct a pH-stat stirred-flow reactor by inserting a micro-combination pH electrode and microburette tip into the reactor. O-rings can provide leak-proof seals where the electrode and burette tip enter the reactor. The electrode and burette tip are connected to an autotitrator as in the conventional batch reactor.

Seyfried et al. (1989) showed that the Carski and Sparks (1985) stirred-flow reactor is a well-mixed system in the flow velocity range of 0.28 to 2.20 mL min^{-1}. A requirement of the stirred-flow method is that the time-dependent effluent solute concentration curves for any kinetic reaction be significantly different from the dilution curve (time-dependent concentration curve with no solid phase in the reactor) and from the time-dependent effluent solute concentration curve for an instantaneous reaction (Sparks, 1989). Although flow velocity can be increased to distinguish fast reactions from instantaneous ones, practical limits on the rates of reactions that can be distinguished do exist. The fastest reaction that can be measured is that producing a time-dependent concentration curve detectably different from the instantaneous case at the flow velocity and reactor volume conditions used. Seyfried et al. (1989) found that reactions with half-lives of greater than 0.3 min could be detected with a flow rate of 0.83 mL min^{-1} in a reactor volume of 8.3 mL. The slowest reaction that can be studied with this method is that producing a time-dependent concentration curve detectably different from the dilution curve (no solid phase in the reactor).

Because flow reactors remove desorbed species, one is tempted to ignore reverse reactions. However, it is only proper to do so when the reaction under consideration is known to be irreversible or when the reverse reaction is negligible compared to the forward one during the observed time frame. As relaxation studies with soil constituents have shown (Hayes and Leckie, 1986; Hachiya et al., 1979; Zhang and Sparks, 1989), reverse reactions can be quite rapid. Thus, the rate of removal of desorbed species in a flow method may not be sufficiently rapid to completely eliminate reverse reactions.

Fluidized Bed Reactor

This type of reactor, which has its origins in chemical engineering, was used by Chou and Wollast (1984) to study albite weathering, but it can be used to study other types of reactions as well. The basic concept behind a fluidized bed is that the flow rate of the fluid is adjusted to equal the settling rate of the particles in suspension. Settling rates of different sized particles tend to be equalized by frequent collisions with other particles if the suspension density is great enough. It is best to use well-defined size fractions to achieve this. A homogeneous, rapidly mixed suspension can be achieved with this method.

One of the advantages of this method is that vigorous mixing in the fluidized bed eliminates strong concentration gradients. Additionally, solute concentrations can be maintained well below saturation levels for various precipitates and thus secondary precipitation reactions that complicate data analysis and interpretation can be avoided. The effect of reaction conditions (e.g., pH changes) can be studied by changing the input solution composition without manipulating the solid phase.

Column Method with Batch Control

This is a hybrid method that combines features of the batch reactor and flow through a soil column to obtain kinetic data. The method was developed by van Riemsdijk and van der Linden (1984) and is an advanced version of the P-stat method of van Riemsdijk and Lyklema (1980a,b). It is particularly suited to the study of the continuing slow reaction between phosphate and soils, where small concentration changes over small time intervals render the thin disk method impractical. In the original P-stat batch reactor method, both pH and P were kept constant because, after a short reaction time, the OH:P ratio of the reaction becomes constant. This approach was particularly useful in studying P retention by oxide surfaces. However, with soils, the OH:P ratio changes with different soils and often the pH change during reaction is too small to control accurately. Thus, van Riemsdijk and van der Linden (1984) developed a new P-stat method applicable to the study of P retention by soils.

Radiotracer Methods

As with the batch reactor method, radiotracers are an excellent means of following the extent of kinetic reactions in flow methods. Flow methods using radiotracers are identical to those where other analytical methods are used to determine the solute of interest except that a radiolabeled solute is used. Radiotracers have been used numerous times in column transport studies, but apparently have not been used much with thin disk and stirred-flow methods.

Advantages and Limitations

Many of the advantages and limitations of flow methods have already been discussed and are summarized here. Advantages include:

1. Low solution:soil ratios more realistically mimic those found under field conditions.
2. Desorbed species are continually removed from the reactor. Their presence can inhibit reaction completion and can sometimes result in complex secondary reactions that complicate data analysis and interpretation.
3. Phase separation is continual and if the flow rate remains constant, separation times remain constant.
4. Flow methods are much more readily automated than batch methods. This reduces operator dependence and error.
5. Diffusion processes are reduced or eliminated if stirred-flow or fluidized bed reactors are used.
6. Direct control with feedback of some reaction conditions can be obtained with some methods such as the stirred-flow or the column method of van Riemsdijk and van der Linden (1984).
7. Reactions can be studied under constant and controlled solute concentrations with some methods.
8. Input solution composition is readily changed with some methods.

Some flow methods, however, do have limitations, including:

1. Reaction rates are usually diffusion limiting in thin disk and column methods.
2. Dilution and washout effects occur with thin disk and column methods.
3. Direct control over reactor conditions is not possible with thin disk methods.
4. In the stirred-flow method, time-dependent concentration curves for reactions must be distinguishable from the instantaneous reaction case and from the dilution curve obtained with no solid phase present.

It is clear that most of the limitations with flow methods apply to the thin disk method. Hybrid methods such as the stirred-flow and fluidized bed reactor combine the best features of batch and flow methods and eliminate or control many of the limitations of each. Future progress in the study of reaction kinetics in soils and soil constituents will most likely come from the use of hybrid batch–flow methods and from the use of relaxation methods where very rapid chemical reactions can be studied.

Comparison of Batch and Flow Methods

Few direct comparisons of batch and flow methods have been done. The most notable are those by Sparks and Rechcigl (1982), Ogwada and Sparks (1986a,b,c), and Miller et al. (1989). Available evidence indicates that there are differences in time to attain equilibrium, half-times of reactions, apparent rate coefficients, and activation energies between batch and thin disk methods as expected. These differences are mostly due to differences in mass transfer rates between the methods. The work of Ogwada and Sparks (1986a,b,c) clearly established the importance of mixing rates and type of agitation in soil kinetic studies. When mass transfer processes control reaction rates and the rate coefficients obtained are those for the overall kinetic process and not those for elementary chemical reactions, then these apparent or process rate coefficients cannot be used to calculate equilibrium constants and thermochemical parameters (Ogwada and Sparks, 1986a).

Equally disturbing is that sorption parameters (i.e., Langmuir sorption parameters) obtained from batch and thin disk methods are different (Miller et al., 1989). This was attributed to the fact that batch systems are closed, whereas flow systems are open so that the competitive antecedent solute species is removed. The importance of the effects of preadsorbed ions on sorption isotherms was well demonstrated by Grolimmund et al. (1995) in flow-through reactor studies. It is apparent that much more work needs to be done on methods development and comparisons. If calculated kinetic parameters are method dependent, then this will obviously limit their usefulness in predicting and understanding solute reactions and transport in soils.

Special Methods for Studying Desorption Reactions

The methods described above can be used to study both sorption and desorption kinetics although the batch method does not readily lend itself to the study of desorption kinetics unless dilution or infinite sink techniques are used. Since desorption studies are approached differently than sorption studies, special techniques are often required and these are discussed next.

Flow Methods

Flow methods, because they readily remove desorbed species, comprise one of the best methods for studying desorption kinetics. Sparks et al. (1980) used the miscible displacement method to study K desorption kinetics. However, mass transfer processes are clearly rate limiting with this method and thus great care must be used in interpreting the results. The previous discussion on the limitations of this method is relevant to desorption studies. Stirred-flow or fluidized bed methods can be used to minimize diffusion in desorption experiments, but they have not yet been used specifically for desorption studies.

Dilution Methods

Dilution of the equilibrium or steady-state solution in contact with the soil has been used to determine the reversibility of sorption reactions (Elrashidi and O'Connor, 1982a, b; Peek and Volk, 1985). The method is also applicable to kinetic studies (Amacher et al., 1986, 1988). The method is particularly suited to sorption and kinetic experiments conducted in centrifuge tubes (Figure 2-5), and the sequence of steps used by Amacher et al. (1986) to obtain their desorption data is shown in Figure 2-7. When the sorption reaction has reached apparent equilibrium or a steady state, a portion (or all) of the solution in contact with the soil is replaced by a solution of identical composition to the equilibrating solution except that the replacement solution does not contain the solute of interest. This results in dilution of the solute concentration in the remaining equilibrating solution and initiates the reverse desorption reaction. Further step-wise dilutions can be done.

The dilution method is not suitable for use with Zasoski and Burau (1978) type batch reactors, since the solid and solution phases must be separated, so that a fraction of the solution phase can be replaced, and then the phases are remixed. Flow methods are of course readily suited for use with the dilution technique, since the input solution can be easily replaced with an identical solution without the solute of interest and then desorption occurs during continuous flow.

Although it has apparently not been used in this regard, the dilution method can be used as a relaxation method. The dilution step serves as the perturbation of the equilibrium or steady state, and the time required to reattain equilibrium is the relaxation time.

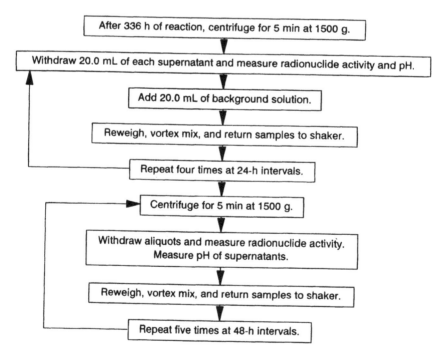

Figure 2-7 Flow chart for batch release studies in centrifuge tubes using radionuclide tracers.

Infinite Sink Methods

Infinite sink methods were developed to overcome problems associated with the accumulation of desorbed species in batch experiments. Infinite sinks are ideally suited to the study of desorption kinetics in batch systems. A sink is a solid phase that removes the desorbing species from solution so that a chemical potential gradient is established. Ideally, a sink should be infinite so that an essentially irreversible reaction from the soil to the sink is established (resorption by soil can be neglected and the rate of sorption of the solute from solution by the sink is much greater than the rate of release from the soil so that solute concentration in solution is minimized). The sorption capacity of the sink must be large enough so that it does not become rate limiting. A variety of infinite sinks have been employed, including ion exchange resins, strongly sorbing mineral phases, and precipitation sinks. Unfortunately, in practice, sinks are not always found to be infinite.

Infinite sinks have been used primarily to study P and K desorption kinetics from soils. Ion exchange resins have been the more popular choice. Recent examples include the studies of P desorption kinetics by Pavlatou and Polyzopoulos (1988) and K desorption kinetics by Sadusky et al. (1987). Yang et al. (1991) used ion exchange resin capsules to desorb nutrients from soils and developed the technology into a new soil test. Cooperband and Logan (1994) used ion exchange resin membranes to measure changes in labile P in soil. van der Zee et al. (1987) used iron oxide-impregnated filter paper as an infinite sink to study P desorption kinetics from soils. The affinity and capacity of this sink for P was large enough to maintain negligible P concentrations in solution and thus it served as an infinite sink. Griffin and Burau (1974) used mannitol as a precipitation sink for boron to study boron desorption from soils. Two separate pseudo-first-order reactions and a very slow reaction were found.

Special Methods for Separating Reactions

There is considerable overlap in reaction times for various reactions in soil systems (Figure 2-1). Thus, separating the various complex reactions to study their kinetics independently is difficult. Some separation is achieved by the simple choice of method. For example, relaxation methods are used to study the kinetics of ion exchange and surface complexation reactions independent from any diffusion processes (Sparks, 1989). Harter and Lehmann (1983) discussed the use of kinetics to distinguish between the almost instantaneous initial ion exchange reaction at soil surfaces and concurrent or consecutive slower ion retention reactions. Jardine and Sparks (1984) used cetyltrimethylammonium bromide to block external ion exchange sites and confirmed that there were two K exchange reactions occurring on an Evesboro soil corresponding to sites of different selectivity. Ogwada and Sparks (1986a,b,c) used different methods primarily based on mixing technique and rate to delineate and separate different diffusion processes, as already discussed. Pavlatou and Polyzopoulos (1988) used the Aharoni and

Suzin (1982a,b) diffusion model to separate time-dependent P desorption curves into regions that could be sequentially described by parabolic, Elovich, and exponential equations at small, intermediate, and large times, respectively. Further discussion of this approach was presented by Aharoni and Sparks (1991).

The initial fast P retention reaction with soils is usually regarded as a surface complexation reaction that can be described with a Langmuir equation, whereas the continuing slow reaction is regarded as precipitation and is usually described by first-order kinetics (Sposito, 1986). However, given the complexity of the processes involved, there is no clear consensus. Mendoza and Barrow (1987) proposed that the continuing reaction between P and soils is the penetration of sorbed P into the sorbing surface (internal diffusion). van der Zee et al. (1989) developed a model to describe both the initial P sorption reaction and the continuing slow reaction. A kinetic Langmuir equation was used to describe adsorption and a diffusion-precipitation model was used to describe the slow reaction. van der Zee et al. (1989) used the iron oxide impregnated filter paper method (van der Zee et al., 1987) to measure P desorption kinetics, a column leaching experiment to establish an adsorption isotherm, and total sorption (adsorption + precipitation) was measured by the van Riemsdijk and van der Linden (1984) method previously discussed. Using these methods and their model, van der Zee et al. (1989) were thus able to separate and describe P reactions with soil.

Amacher et al. (1988), Selim and Amacher (1988), and Harter (1989) developed nonlinear, second-order, and first-order multireaction models, respectively, to describe element retention by soils. Reactions are distinguished on the basis of mass-action kinetics and these models were found to describe the kinetics of element retention in soils when single-reaction rate functions failed (Amacher et al., 1986). This approach is the basis for the subsequent chapters in this book.

The approaches of Harter and Lehmann (1983), Aharoni and Suzin (1982a,b), Amacher et al. (1988), Selim and Amacher (1988), and Harter (1989) are model-based approaches in that data analysis methods are used to distinguish reactions. The approach of van der Zee et. al. (1989) is both a model and experimental method-based approach to separating reactions. The other techniques described are primarily experimental. Given the complexity of reactions in soils, it is clear that further work on reaction separation methods are needed.

DATA ANALYSIS METHODS

Once the kinetic data is obtained, it must be analyzed. The objective is to obtain an overall rate function that will describe the observed reaction kinetics. Determination of the rate function also involves determination of the reaction order and rate coefficients. A number of useful methods are available to derive overall rate functions from the experimental data, and each of these is described in turn.

Initial Rate Method

This method depends on measurement of initial rates of an overall reaction before effects of accumulating products (desorbed species) and effects of decreasing reactant concentrations complicate the rate function. The method depends on the use of sensitive analytical methods (e.g., radiotracer methods) to determine rates at small reaction extents. If the overall rate is sufficiently slow such that no significant chemical changes occur during the measurements, the initial amounts of each reactant can be varied while holding the others constant to find the relationship between initial rate and initial reactant concentration. This relationship is given by

$$\log R_i = \log k' + n_A \log C_A^o \qquad (2\text{-}20)$$

where R_i is the initial rate, k' is the pseudo rate coefficient, n_A is the reaction order for reactant A, and C_A^o is the initial concentration of A.

A plot of R_i vs. C_A^o will be a straight line with slope n_A and intercept k'. This equation is applied to each reactant until the reaction orders for all reactants are found (Lasaga, 1981; Gardiner, 1969).

The initial rate method has rarely been applied to soil kinetics studies. Aringhieri et al. (1985) used the initial rate method to establish that the reaction orders for each reactant (element and soil) were unity for Cu and Cd retention by a soil. The overall reaction was thus second order. Rimstidt and Newcomb (1993) presented a general differential method for analyzing rate data for reactions far from equilibrium in which there is no significant reverse reaction.

The major obstacle to the use of the initial rate method to determine reaction orders is that most ion exchange and surface complexation reactions occur so rapidly that the initial rate cannot be accurately determined before the reverse reaction becomes significant. Flow methods with high flow velocities perhaps offer the best chance of measuring initial rates of some of the slower reactions because desorbed species are quickly transported away from soil surfaces.

Method of Isolation

In the method of isolation, all reactants except one are present at large concentrations during the reaction. The rate function for the "isolated reactant" is then determined. This process is repeated for all reactants (Gardiner, 1969; Lasaga, 1981). For example, if the true overall rate function for a reaction is:

$$\frac{dC_A}{dt} = -kC_A C_B \qquad (2\text{-}21)$$

and if the initial concentration of B, C_B^o, is much greater than C_A^o so that it does not vary significantly over the course of the reaction, then the rate function can be rewritten as:

$$\frac{dC_A}{dt} = -kC_B^o C_A = -k'C_A \qquad (2\text{-}22)$$

where $k' = kC_B^o$.

Thus, the second-order rate function reduces to a first-order function.

The method of isolation is often assumed in kinetic studies on soils and soil constituents. Frequently, first-order kinetics for the solute of interest is assumed but never tested, and the reaction sites on the solid phase are assumed to be present in excess of the solute so that the reaction rate does not depend on their concentration. This assumption is probably erroneous. Often the quantity of reaction sites on the solid is not known with any degree of certainty. In the case of exchangeable cations, the total quantity of reaction sites is given by the cation exchange capacity (CEC). In the case of other solutes like transition metals and oxyanion species that can react with metal oxide and organic matter surfaces with variable charge, the total quantity of reaction sites is not known with certainty. It can be estimated by the maximum sorption capacity as determined by the family of Langmuir equations applied to solute sorption data. The real complicating factor, however, is the possible presence of different types of reaction sites in soils that display different reaction kinetics. When this occurs, the quantities of the different types of reaction sites are not known and can often only be guessed. Although the total quantity of reaction sites may be in excess, quantities of some types may be limiting and thus pseudo-first-order kinetics is not obeyed. This situation may occur more often than is realized (Amacher et al., 1988; Selim and Amacher, 1988).

Graphical Methods

Often, the fit of the data to an assumed integrated rate function is used as a test of the validity of that rate function (Gardiner, 1969; Lasaga, 1981). Many integrated rate functions are linear functions of time so that if the data are plotted according to the integrated rate function and a straight line is obtained, then the data is said to follow that rate function. One of the simplest cases is if the data gives a straight line on a plot of reactant concentration on a log scale vs. time, then first-order kinetics is obviously obeyed. Unfortunately, soil kinetic data seldom show correspondence to simple integrated rate functions (Amacher et al., 1986).

Rate Coefficient Constancy

Assumed integrated rate functions are again used in this approach. Rate coefficients are calculated for each assumed rate function applied to data sets from kinetic experiments where variables such as initial reactant concentrations were systematically varied. The rate coefficients are then compared to each other.

The correct rate function should be the one for which a set of calculated rate coefficients shows only random scatter about the average value for the set. Incorrect rate functions will have rate coefficients that vary systematically with initial concentration and time (Gardiner, 1969). The latter situation is frequently observed for soil kinetic data (Amacher et al., 1988).

Fractional Lives Method

In this method, the time required for a given fractional decrease in starting concentration (frequently the half-time of the reaction) is measured as a function of initial reactant concentrations. A log–log plot of fractional lives vs. initial concentration will give the reaction order (Gardiner, 1969). Often, fractional lives methods will be used as a convenient way of analyzing kinetic data. Boyd et al. (1947) and Kressman and Kitchener (1949) used a fractional lives type approach to analyze their ion exchange kinetic data. A reaction half-time approach to analyzing soil kinetic data was discussed by Seyfried et al. (1989).

Parameter-Optimization Methods

In this approach, a particular rate function is assumed and nonlinear least-squares parameter-optimization techniques are used to calculate rate coefficients. Many techniques are available and a computer program developed by Parker and van Genuchten (1984) is good for this purpose. It is basically the maximum neighborhood method of Marquardt (1963). Various statistics are used to evaluate goodness-of-fit of the rate functions to the data including r-square, root mean square, 95% confidence intervals for computed parameters, and the parameter correlation matrix. The rate function(s) that give the best fit to the data are then assumed to be the most nearly correct.

Integrated forms of the simple mass action rate functions produce linear equations that are easily tested by graphical methods. The advantage of parameter-optimization methods is that the computer programs can be written to generate statistics for a more quantitative estimation of goodness-of-fit rather than the visual estimation that graphical methods provide.

Numerous comparisons of rate functions for describing soil kinetic data have been done (e.g., Onken and Matheson, 1982; Sparks and Jardine, 1984; Havlin et al., 1985; Amacher et al., 1986). It is often found that simple, single, reaction-rate functions do not adequately describe the data or that more than one rate function will describe the data equally well. Possible reasons for this are that more than one reaction is occurring, that mass transfer and reaction processes are occurring together (elementary reactions are not being measured), or that the kinetics are more complex than the assumed rate function.

The failure of single reaction rate functions to adequately describe soil kinetic data led Amacher et al. (1988) and Selim and Amacher (1988) to develop nonlinear and second-order multireaction rate functions that form the basis for the remaining chapters in this book. A similar approach was used by Harter (1989), where up to three concurrent reversible first-order rate functions were considered. Many alternative multireaction rate functions will produce time-dependent concentration curves that are statistically indistinguishable. This point was discussed thoroughly by Skopp (1986) in his review of kinetic processes in soils and was amplified by Amacher et al. (1988), Selim and Amacher (1988), and Harter (1989). In such cases where more than one explanation for the observed kinetics is possible, experimental evidence must be used to support a particular mechanism. Curve fitting alone will not suffice.

When using multireaction rate functions to describe kinetic data, the simplest model with the best overall fit to the data and which has the lowest parameter standard errors is the most desirable. One must guard against overfitting the data, which occurs by using a model with too many parameters for too few data points. The extra sum of squares principle (Kinniburgh, 1986) can be used to determine if there is any statistically significant improvement in the fit of the model to the data by adding more parameters (i.e., more reactions). Inflated parameter standard errors are an indication of an incorrect model choice or too many parameters. A strong linear dependence among model parameters will also occur if the model overfits the data. This can be evaluated by examining the parameter correlation matrix.

Kinetic equations (Table 1-1) can be used to directly analyze batch kinetic data since no transport (flow) processes occur. In flow methods, the appropriate transport equation must be coupled to the kinetic equation to achieve a correct solution. For the thin disk method, the transport equation (Skopp and McCallister, 1986) is

$$V\Theta \frac{\partial C}{\partial t} = AJ(C_{in} - C_{out}) - V\rho \frac{\partial S}{\partial t} \tag{2-23}$$

where V is the thin disk reactor volume (cm^3), Θ is the volumetric water content (cm^3 cm^{-3}), C is the solute concentration in the reactor (mol L^{-1}), A is the cross-sectional area of the thin disk (cm^2), J is the flow rate or flux (cm^3 sec^{-1}), C_{in} and C_{out} are the influent and effluent solute concentrations, respectively (mol L^{-1}), ρ is the bulk density of the solid on the thin disk (g cm^{-3}), and S is the concentration of solute on the soil (mol kg^{-1}).

Analytical solutions are available for simple kinetic equations (e.g., Skopp and McCallister, 1986). Numerical approximations can be used for more complex kinetic equations.

A different transport equation (Schnabel and Fitting, 1988) is used for stirred-flow reactors:

$$V\frac{\partial C}{\partial t} = J(C_{in} - C) - m\frac{\partial S}{\partial t} \tag{2-24}$$

where C is the solute concentration in the reactor and in the effluent, m is the mass of soil in the reactor, and the other terms are defined as before.

Schnabel and Fitting (1988) give a solution for the case of first-order kinetics. Numerical approximation techniques can be used for more complex kinetic equations, but the use of multireaction models in flow systems has not yet become a reality.

EFFECTS OF EXPERIMENTAL VARIABLES ON RATE FUNCTIONS

A kinetic study is not complete unless the effects of various experimental variables on the experimental rate functions are determined. By systematically changing experimental variables and determining the effect on the rate function, valuable clues are obtained that will aid in deducing mechanisms to explain the observed rate function. The most commonly manipulated variables in soil kinetic studies include reactant concentrations (both solute and solid phase), temperature, pH, ionic strength, and solution composition (other than pH, ionic strength, and solute concentration).

Reactant Concentrations

Obtaining kinetic data over a wide range of initial reactant concentrations is essential to determine unambiguously the correct rate function to describe the data. Unfortunately, systematic variation of initial reactant concentrations is not often practiced in soil kinetic studies. The initial concentration of the solute reactant is the one most often varied and this is easy to do in both batch and flow methods. Varying the concentration of the solid phase (solution:soil ratio) is not possible in the thin disk method and can only be done over a narrow range in stirred-flow methods. Wide variations in solution:soil ratios are generally possible only with batch methods. Changing the amount of soil per given volume of solution also changes the amount of reaction sites. It is not unusual to discover that rate coefficients calculated from soil kinetic data using an assumed rate function vary systematically with initial reactant concentrations, indicating that the reaction is more complex than that implied by the assumed rate function (Amacher et al., 1988).

Temperature

This is one of the most important variables to study and temperature effects in soil kinetic studies have been frequently reported (e.g., Sparks and Jardine, 1981; Ogwada and Sparks, 1986a; Barrow and Shaw, 1975; Evans and Jurinak, 1976). Rate coefficients for elementary chemical reactions follow a temperature dependence that can be described by the well-known Arrhenius equation:

$$k = A \exp^{-E/RT} \tag{2-25}$$

where A is the pre-exponential factor, E is the activation energy, R is the universal gas constant, and T is absolute temperature.

The magnitude of the activation energy is often used as a criteria to distinguish between diffusion-controlled and reaction-controlled kinetics (Sparks, 1985, 1986, 1989). Low activation energies are indicative of diffusion-controlled kinetics, whereas high activation energies are indicative of a chemical reaction. This is logical since chemical bonds are broken and formed in chemical reactions and the energy barrier that must be overcome can be quite high. Diffusion on the other hand requires no chemical bond breaking or formation, so energy barriers are quite low. Boyd et al. (1947) and Kressman and Kitchener (1949) used temperature effect criteria to support the diffusion-controlled kinetic mechanism in their now classic ion exchange studies. Ogwada and Sparks (1986a,b,c) clearly demonstrated the effect of mixing on activation energy values for ion exchange. Under static or low agitation conditions where diffusion-limiting kinetics occur, activation energies were low, but increased sharply under vigorous mixing conditions where diffusion was no longer limiting.

pH

The effects of pH on sorption isotherms have been studied extensively, particularly with oxide surfaces (Anderson and Rubin, 1981; Sposito, 1984), but pH effects in sorption kinetic studies have not received equal attention. In contrast, pH effects in mineral dissolution kinetic experiments have received a great deal of attention (e.g., Stumm, 1986; Chou and Wollast, 1984; Stone, 1987a,b).

Ionic Strength

Transition-state theory (Lasaga, 1981; Gardiner, 1969) predicts that rate coefficients for second-order reactions in solution depend on the activity coefficients of the reactants and activated complex and therefore vary with ionic strength (the primary salt effect); this has been found to be the case. However, the dependence of rate coefficients of kinetic reactions in soils on ionic strength has apparently not been studied, and the dependence of rate coefficients in soils on ionic strength has received little attention.

Solution Composition

The effects of solution composition (aside from reactant concentrations, pH, and background electrolyte concentrations) on reaction kinetics should also be studied. The use of different but chemically related reactants and competitive effects from other species in solution are examples of this approach. For example, Stone and Morgan (1984) examined the effect of different organic compounds on dissolution rates of manganese oxides.

COLUMN TRANSPORT METHODS

One of the objectives of obtaining rate coefficients from kinetic studies is to use them in predicting reactive solute transport in soils. In this section, we describe a laboratory method for obtaining breakthrough curves (BTCs) for solute transport during saturated flow-through soil columns of finite length. This method is suitable for obtaining experimental BTCs that can be used to test various models for describing reactive solute transport in soils. Such an approach is a desirable prerequisite for model evaluation before addressing more complex reactive solute transport cases such as unsaturated flow and physical and chemical heterogeneity encountered in field situations.

Figure 2.8 shows a diagram of the experimental setup that can be used to obtain BTCs for nonreactive and reactive solutes. A reservoir containing the solute of interest in a background electrolyte solution, pump, soil column, and fraction collector are connected in series.

The pump must be capable of maintaining a constant flow velocity for the duration of each solute transport experiment. It is desirable to use a pump capable of a range of flow velocities, so that the effect of flow rate on the experimental BTCs can be examined. A piston-type fluid metering pump works well for solute transport studies.

The soil column is typically constructed of plexiglass and has end caps with screw threads and O-rings to ensure a leak-proof seal during the experiment. Hose-nipple connections in the center of each end cap make connection of the tubing easy. Some type of filter (quantitative filter paper, glass fiber filter, membrane filter, or porous plate) is placed at each end of the soil column to contain the soil. Column sizes vary, but columns of 4.5 to 7.5 cm in diameter and 5 to

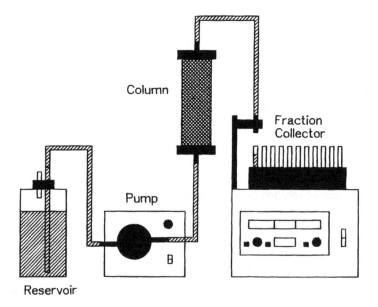

Figure 2-8 Solute transport experimental setup. Background solution and solute are pumped from the reservoir through the soil column and are collected as aliquots by the fraction collector. Separation of solid and aqueous phases is accomplished by a filter at the outlet end of the soil column.

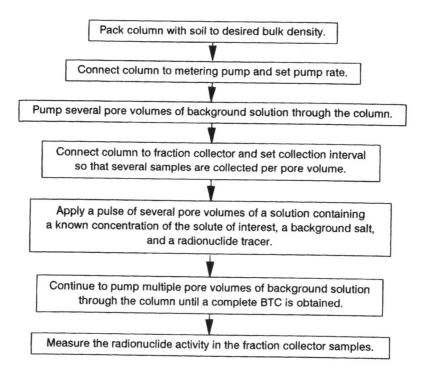

Figure 2-9 Flow diagram for conducting a solute transport experiment through a soil column.

30 cm in length are common. To avoid overly long times until solute breakthrough, the size of the column should be selected based on the reactivity of the solute with the soil of interest at the flow rate to be used. A highly reactive solute moving through a soil with a high capacity for sorption of that solute will require a long breakthrough time. A shorter soil column will avoid overly long experiments. Batch experiments provide valuable information on reactivity of the solute and sorption capacity of the soil. A fraction collector set to collect aliquots of column effluent at timed intervals completes the experimental setup.

A flow chart of a typical solute transport experiment is shown in Figure 2-9. Typically, the soil in the column is intact or undisturbed representing *in situ* soil condition. Alternatively, the soil must be packed uniformly to a known bulk density computed from the weight of the soil in the known volume of the soil column. The soil bulk density is an input parameter needed for the models described in subsequent chapters. The volumetric water content is another input parameter needed for the models and may be directly measured or estimated from the pore volume of the packed soil column:

$$\Theta = \frac{V_P}{V_T} \tag{2-26}$$

$$V_P = \left(1 - \frac{\rho_b}{\rho_p}\right)V_T \qquad (2\text{-}27)$$

where Θ is the volumetric water content (cm^3 cm^{-3}), V_P is the pore volume (cm^3), V_T is the total volume of the column (cm^3), ρ_b is the soil bulk density (g cm^{-3}), and ρ_p is the particle density (assumed to be 2.65 g cm^{-3}).

Before applying a pulse of solute to the soil column, a background electrolyte solution [e.g., 0.01 M $CaCl_2$ or $Ca(NO_3)_2$] is often pumped through the column to fill the pore volume and saturate exchange sites with the cation and anion of the background solution. A background solution is also used to maintain the ionic strength of the aqueous phase at a constant level. During the time the background electrolyte is being pumped through the column, the flow rate is adjusted to the desired level.

A pulse of a nonreactive solute (e.g., tritium, chloride) should be applied to the soil column, followed by background solution to obtain a complete BTC for the nonreactive solute. The BTC for the nonreactive solute is needed to determine the dispersion coefficient for the convective-dispersive transport equation (Chapter 4). The nonreactive solute BTC will also provide valuable information on the effect of physical heterogeneity of the soil on solute transport and whether a two-region (mobile–immobile water) transport model may be more appropriate for the soil being studied (Chapter 6).

To begin the actual solute transport experiment, the inlet tubing is placed in a reservoir of background solution containing a known concentration of the solute of interest. Several pore volumes of solute and background solution are pumped through the soil column (solute pulse), followed by multiple pore volumes of just background solution to produce the desorption part of the BTC. The amount of the solute input pulse (concentration $\times$ number of pore volumes) and the number of pore volumes of background solution to obtain a complete BTC will depend on the reactivity of the solute with soil. A relatively nonreactive solute-soil combination requires a smaller solute pulse and fewer pore volumes of background solution to achieve a complete BTC than a more reactive combination. Batch sorption data will provide valuable information on the concentration and number of pore volumes of the solute pulse to apply to the soil as well as how long the column must be eluted with background solution to produce the desorption side of the BTC.

Column effluent samples are conveniently collected with a fraction collector set at a constant time interval. Several samples per pore volume should be collected. Samples can be analyzed for the solute of interest. For many solutes, transport experiments can be performed using a radionuclide tracer of the solute of interest added to the solute and background solution. The radionuclide activity of the fraction collector samples is then counted to determine the concentrations of solute in the samples. This approach was used by Selim and Amacher (1988) and Selim et al. (1989) for their Cr(VI) transport studies.

Breakthrough curves are most conveniently presented in terms of dimensionless units. Reduced concentration values (C/C_o) are plotted against reduced time [number of pore volumes (V/V_o)] to produce the BTC.

For some studies it may be desirable to section the soil column after the solute transport experiment to determine the distribution of solute within the soil using various instrumental or chemical analysis techniques. Such analyses provide valuable information on the types of reactions that should be included in the solute transport model.

3 KINETIC RETENTION APPROACHES

For several heavy metals (e.g., Cr, Cu, Zn, Cd, and Hg), retention/release reactions in soils have been observed to be strongly time-dependent. Studies on the kinetic behavior of several heavy metals in soils include those by Harter and Lehmann (1983), Aringhieri et al. (1985), and Amacher et al. (1986), among others. A number of empirical models have been proposed to describe kinetic retention/release reactions of solutes between the solution and solid phases. The earliest model is the first-order kinetic equation first incorporated into the convection–dispersion transport equation by Lapidus and Amundson (1952). First-order kinetic models have been extended to include the nonlinear kinetic type (van Genuchten et al., 1974, Mansell et al., 1977, Fiskell et al., 1979). A variety of other kinetic reactions are given by Murali and Aylmore (1983).

Amacher et al. (1986) found that the use of single-reaction kinetic models did not adequately describe the time-dependent retention of Cr, Cd, and Hg for several initial (input) concentrations and several soils. As a result, Amacher et al. (1988) developed a multireaction model that includes concurrent and concurrent–consecutive processes of the nonlinear kinetic type. The model was capable of describing the retention behavior of Cr(VI) and Cd with time for several soils. In addition, the model predicted that a fraction of these heavy metals was irreversibly retained by the soil.

Selected examples of the kinetic retention of Cd by several soils in batch experiments are given in Figure 3-1 (Selim, 1989). The amount of Cd retained varied among the soils, with Cecil soil exhibiting the lowest retention and Sharkey soil showing maximum Cd sorption from soil solution. The rapid initial decrease in Cd concentration with time indicates a fast initial sorption reaction, followed by slower sorption reactions. It is also apparent that after 300 h of reaction time, quasi-equilibrium conditions were not yet attained in several of the soils. The slow approach to equilibrium is further illustrated in Figure 3-2, which shows sorption isotherms for Cd on Windsor soil at several reaction times (Selim et al., 1992). It is obvious that when dealing with such data sets, the use of equilibrium-type models will provide inadequate predictions of the fate of metals in these soils.

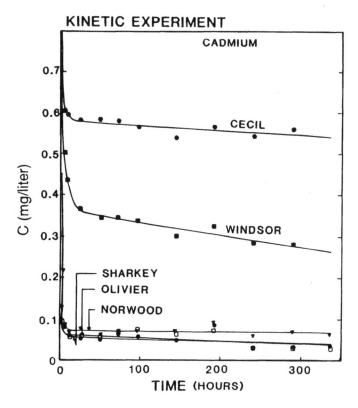

Figure 3-1 Time-dependent retention of Cd by five soils at an initial Cd concentration (C_O) of 1 mg L^{-1}.

SINGLE-REACTION MODELS

A number of empirical models have been proposed to describe kinetic retention reactions of solutes in the solution phase. The earliest model is the first-order kinetic reaction first incorporated into the classical (convective-dispersive) transport equation by Lapidus and Amundson (1952):

$$\rho \frac{\partial S}{\partial t} = k_f \Theta C - k_b \rho S \qquad (3\text{-}1)$$

where S is the amount of solute retained by the soil (mg kg^{-1}), C is the solute concentration in solution (mg L^{-1}), k_f and k_b are the forward and backward reaction rate coefficients (h^{-1}), respectively, ρ is the soil bulk density (g cm^{-3}), and Θ is the soil water content (cm^3 cm^{-3}).

This type of reaction is fully reversible where the magnitudes of the rate coefficients dictate the extent of the kinetic behavior of the reaction. For small values of k_f and k_b, the rate of retention is slow and strong kinetic dependence is anticipated. In contrast, for large values of k_f and k_b, the retention reaction is

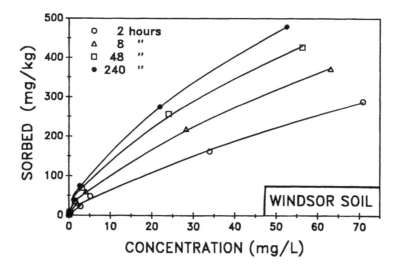

Figure 3-2 Sorption isotherms for Cd retention by Windsor soil at different times of retention. Solid curves are predictions using the Freundlich equation.

rapid and should approach quasi-equilibrium in a relatively short time. In fact, at large times ($t \rightarrow \infty$), when the rate of retention approaches zero, the above equation yields:

$$S = \left(\frac{\Theta}{\rho}\right) \frac{k_f}{k_b} = K_d C \qquad (3\text{-}2)$$

which results in a linear equation where $K_d = \Theta k_f / \rho k_b$ (cm^3 g^{-1}). This is similar to that for linear isotherms where equilibrium conditions were assumed. In addition, as discussed earlier, nonlinear retention has been commonly observed for several solutes (see also Figure 3.3). As a result, the first-order reversible kinetic reaction has been extended to include the nonlinear kinetic type (Mansell et al., 1977):

$$\rho \frac{\partial S}{\partial t} = k_f \Theta C^n - k_b \rho S \qquad (3\text{-}3)$$

where n is a dimensionless parameter commonly less than unity and represents the order of the nonlinear reaction.

In a similar formulation to the linear case, at large times ($t \rightarrow \infty$), when the rate of retention approaches zero, the above equation reduces to:

$$S = K_f C^n \qquad (3\text{-}4a)$$

where

$$K_f = \left(\frac{\Theta}{\rho}\right)\frac{k_f}{k_b} \qquad (3\text{-}4b)$$

which is commonly known as the Freundlich equation. Therefore, for linear or Freundlich isotherms, one may regard the distribution coefficients, K_d and K_f, as the ratio of the rate of sorption (forward reaction) to that for desorption or release (backward reaction).

To illustrate solute retention behavior (adsorption–desorption) when the first-order or the nonlinear kinetic equations are used, we present several simulations in Figures 3-3 and 3-4 (Selim et al., 1976). As shown in Figure 3-3 (top), the linear kinetic adsorption isotherms are strongly time-dependent and, even after more than 50 h, only 90% of equilibrium was achieved. The slow attainment of equilibrium can be attributed primarily to the magnitude of the reaction rate

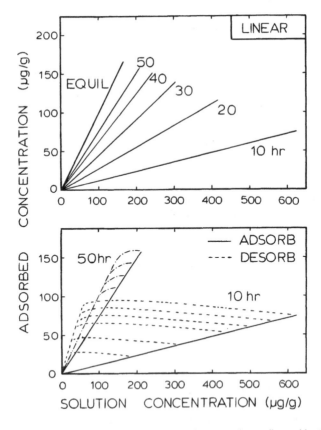

Figure 3-3 Simulated adsorption–desorption isotherms using a linear kinetic retention model. Desorption was initiated after 10 and 50 h for each successive sorption step.

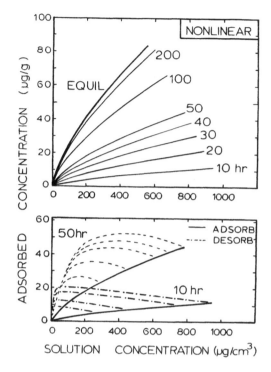

Figure 3-4 Simulated adsorption–desorption using a nonlinear kinetic retention model. Desorption was initiated after 10 and 50 h for each successive sorption step.

coefficients k_f and k_b. Figure 3-3 (bottom) also shows simulated adsorption curves for 10 and 50 h and desorption (dashed) curves initiated after 10 and 50 h of adsorption. The simulated desorption isotherms shown were obtained by successively reducing the solute solution concentration by one half every 10 h (or 50 h) until the solution concentration was less than 5 mg L^{-1}. This procedure is similar to that used in desorption studies in the laboratory. As seen from the family of (dashed) curves, desorption did not follow the same path (i.e., nonsingularity) as the respective adsorption isotherm (solid curves). Obviously, this nonsingularity or hysteresis results from the failure to achieve equilibrium adsorption prior to desorption. If adsorption as well as desorption were carried out for times sufficient for equilibrium to be attained, or the kinetic rate coefficients were sufficiently large, such hysteretic behavior would be minimized.

The time-dependent adsorption and the nonsingularity or hysteretic behavior of the adsorption–desorption processes for the case of nonlinear retention (see eq. 3-3) are shown in Figure 3-4. Reaction times greater than 200 h were required to achieve >90% equilibrium in the nonlinear case. It is apparent that the hysteretic property is for linear or nonlinear kinetic reactions. Hysteretic behavior has been observed by several scientists (Munns and Fox, 1976; Flühler, et al., 1982). An example of hysteretic behavior for Cu sorption and desorption is shown in Figure 3-5.

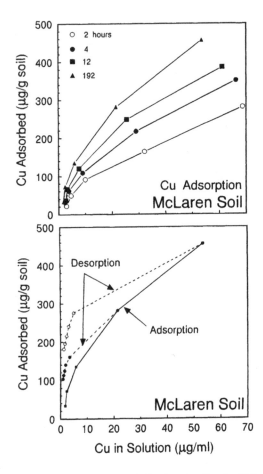

Figure 3-5 Adsorption–desorption hysteresis for Cu retention by McLaren soil.

First-order kinetic reactions have also been used to quantify various irreversible (sink/source) reactions in soils, including precipitation/dissolution, mineralization, immobilization, biological transformations, volatilization, and radioactive decay, among others. Models that include first-order kinetic (sink/source) and sequential first-order (irreversible) decay reactions include those by Cho (1971), Selim and Iskandar (1981), Rasmuson and Neretienks (1981), and Amacher et al. (1988). The first-order irreversible retention form is:

$$Q = k_s \Theta C \tag{3-5}$$

where k_s is the irreversible reaction rate coefficient (h^{-1}).

On the other hand, description of precipitation reactions that involve secondary nucleation is not an easy task, and it is often difficult to distinguish between precipitation and adsorption. In fact, Sposito (1984) stated that the problem of differentiating adsorption from precipitation is made more severe by the facts that new solid phases can precipitate homogeneously onto the surfaces of existing solid phases and that weathering solids may provide host surfaces for the more stable phases into which they transform chemically.

First-order kinetic reactions have also been used as an approximate method for describing the retention reaction due to ion exchange on pure clays. Jardine and Sparks (1984) studied the kinetics of potassium retention on kaolinite, montmorillonite, and vermiculite and found that the adsorption process is kinetic in nature. They also found that a single first-order decay-type reaction described the data adequately for kaolinite and montmorillonite, whereas two first-order reactions were necessary to describe potassium retention on vermiculite. Jardine and Sparks (1984) suggested that deviations of experimental data from first-order kinetics at larger times (when equilibrium is approached) is likely due to the fact that potassium retention is not an irreversible but rather a reversible mechanism. At large times (or large amount of potassium adsorbed), the contribution of the reverse or backward retention process becomes significant and thus should not be ignored. Other cations and anions that exhibited kinetic ion exchange behavior include Al, NH_4, and several heavy metals.

An extensive list of studies illustrating kinetic behavior has been recently compiled by Sparks (1989). According to Ogwada and Sparks (1986), observed kinetic ion exchange behavior in soils is probably due to mass transfer (or diffusion) and chemical kinetic processes. They stated that for chemical sorption to occur, ions must be transported to the active (fixed) sites of the soil particles. The film of water adhering to and surrounding the particles and water within the interlayer spaces of the particles are both zones of low concentrations due to depletion by adsorption of ions onto the exchange sites. The decrease in concentration in these two interface zones may be compensated by diffusion of ions from the bulk solution.

Other kinetic models used to describe the rate of reversible and irreversible reactions include the Elovich model and the diffusion-controlled model. A variety of other kinetic reactions are given by Travis and Etnier (1981) and Murali and Aylmore (1983), and a detailed discussion of the characteristics of several reactions is available (Sparks, 1989).

MULTIPLE-REACTION MODELS

The problem of identifying the fate of solutes in soils must account for retention reactions and transport of the various species in the soil environment (Theis, 1988; Barrow, 1989). In fact, Barrow (1989) stated that the use of single-reaction models, such as those described above, is not adequate since such models describe the fate of only one species, with no consideration to the simultaneous reactions of others in the soil system. This is supported by the work of Amacher et al. (1986) who showed that sorption–desorption of Cr, Cd, and Hg from batch studies on several soils was not described by use of single-reaction models of the equilibrium Langmuir or Freundlich type. They also found that a first-order kinetic reaction was not capable of describing Cr, Cd, and Hg concentrations in the soil solution with contact time. Aringhieri et al. (1985) showed that retention of Cd and Cu on an organic soil was highly kinetic. A description of Cd and Cu reaction using a second-order approach gave adequate predictions of their behavior provided the reaction rate coefficients were time-dependent.

Multisite or multireaction models deal with the multiple interactions of one species in the soil environment. Such models are empirical in nature and are based on the assumption that a fraction of the total sites reacts rapidly or instantaneously with the solute whereas the remaining fraction of sites reacts slowly with solute in the soil solution (Selim et al., 1976; Jardine et al., 1985b). Nonlinear equilibrium (Freundlich) and first- or n^{th}-order kinetic reactions are the associated processes. Such a two-site approach proved successful in describing the observed extensive tailing of solute breakthrough curves. Another two-site approach was proposed by Theis et al. (1988) for Cd mobility and adsorption on goethite. They assumed that sorption reactions with both types of sites were governed by second-order kinetics. The reactions were assumed to be consecutive where the second reaction was irreversible. Amacher et al. (1988) developed a multireaction model that includes concurrent and concurrent-consecutive processes of the nonlinear kinetic type. The model was capable of describing the retention behavior of Cr(VI) and Cd with time for several soils. In addition, the model predicted that a fraction of these heavy metals was irreversibly retained by the soil. Recently, Amacher et al. (1990) concluded that the multireaction model was also successful in describing sorption of Hg by several soils.

Two-Site Models

One of the earliest multireaction models is the two-site model proposed by Selim et al. (1976). This model was developed to describe observed batch results that showed rapid initial retention reactions followed by slower retention reactions. The model was also developed to describe the excessive tailing of breakthrough curves obtained from pulse inputs in miscible displacement experiments. The two-site model is based on several simplifying assumptions. First, it is assumed that a fraction of the total sites (referred to as type I sites) reacts rapidly with the solute in soil solution. In contrast, we assume that type II sites are highly kinetic in nature and react slowly with the soil solution. The retention reactions for both types of sites are based on the nonlinear (or n^{th} order) reversible kinetic approach and may be expressed as:

$$\frac{\partial S_1}{\partial t} = k_1 \frac{\Theta}{\rho} C^n - k_2 S_1 \qquad (3\text{-}6)$$

$$\frac{\partial S_2}{\partial t} = k_3 \frac{\Theta}{\rho} C^m - k_4 S_2 \qquad (3\text{-}7)$$

$$S_T = S_1 + S_2 \tag{3-8}$$

where S_1 and S_2 are the amounts of solute retained by sites I and sites II, respectively, S_T is the total amount of solute retained by the soil matrix (mg kg^{-1}), and k_1, k_2, k_3, and k_4 are the associated rate coefficients (h^{-1}).

The nonlinear parameters n and m are considered to be less than unity and $n \neq m$. For the case $n = m = 1$, the retention reactions are of the first-order type and the problem becomes a linear one.

This two-site approach was also considered for the case where type I sites were assumed to be in equilibrium with the soil solution, whereas type II sites were considered to be of the kinetic type. Such conditions may be attained when the values of k_1 and k_2 are extremely large. Under these conditions, a combination of equilibrium and kinetic retention is (Selim et al., 1976):

$$S_1 = K_f C^n \tag{3-9}$$

$$\frac{\partial S_2}{\partial t} = k_3 \frac{\Theta}{\rho} C^m - k_4 S_2 \tag{3-10}$$

Jardine et al. (1985b) found that the use of the equilibrium and kinetic two-site model provided good predictions of breakthrough curves (BTCs) for Al in kaolinite at different pH values. Selim et al. (1976) found that the two-site model yielded improved predictions of the excessive tailing of the desorption or leaching side and the sharp rise of the sorption side of the BTCs in comparison to predictions using single-reaction equilibrium or kinetic models. The two-site model has been used by several scientists, including DeCamargo et al. (1979), Nkedi-Kizza et al. (1984), Jardine et al. (1985b), and Parker and Jardine (1986), among others. The model proved successful in describing the retention and transport of several dissolved chemicals including Al, P, K, Cr, Cd, 2,4-D, atrazine, and methyl bromide. However, there are some inherent disadvantages to the two-site model. First, the reaction mechanisms are restricted to those that are fully reversible. Moreover, the model does not account for possible consecutive type solute interactions in the soil system.

Multireaction Models

A schematic representation of the multireaction (MRM) model is shown in Figure 3-6. In this model, we consider the solute to be present in the soil solution phase (C) and in five phases representing solute retained by the soil matrix as S_e, S_1, S_2, S_3 and S_{irr}. We further assume that S_e, S_1 and S_2 are in direct contact with the solution phase and are governed by concurrent type reactions. Here we assume S_e is the amount of solute that is sorbed reversibly and is in equilibrium with C at all times. The governing equilibrium retention/release mechanism is that of the nonlinear Freundlich type as discussed previously.

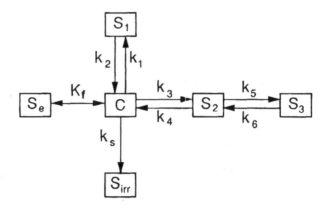

Figure 3-6 A schematic representation of the multireaction model MRM.

The retention/release reactions associated with S_1 and S_2 in direct contact with C are reversible processes of the nonlinear kinetic type:

$$S_e = K_f C^b \tag{3-11}$$

$$\frac{\partial S_1}{\partial t} = k_1 \frac{\Theta}{\rho} C^n - k_2 S_1 \tag{3-12}$$

$$\frac{\partial S_2}{\partial t} = k_3 \frac{\Theta}{\rho} C^m - k_4 S_2 \tag{3-13}$$

These two phases (S_1 and S_2) may be regarded as the amounts sorbed to surfaces of soil particles such as surface complexes of Al and Fe oxide minerals or other types of surfaces, although it is not necessary to have *a priori* knowledge of the exact retention mechanisms for this model to be applicable. Moreover, these phases may be characterized by their kinetic sorption/desorption interaction with the soil solution. Thus, any solute whose retention/release behavior in soils can be described by such a model is susceptible to leaching in the soil. In addition, the primary difference between these two phases not only lies in the difference in their kinetic behavior, but also on the degree of nonlinearity, as indicated by the parameters n and m.

The multireaction model also considers irreversible solute removal via a retention sink term, Q, to account for possible irreversible reactions such as precipitation or internal diffusion among others. We express the sink term as a first-order kinetic process:

$$Q = \rho \frac{\partial S_{irr}}{\partial t} = k_s \Theta C \tag{3-14}$$

where k_s is the associated rate coefficient (h^{-1}).

The multireaction model also includes an additional retention phase (S_3) governed by a consecutive reaction with S_2. This phase represents the amount of solute strongly retained by the soil that reacts slowly and reversibly with S_2 and may be a result of further rearrangements of the solute retained on matrix surfaces. Thus, inclusion of S_3 in the model allows the description of the frequently observed very slow release of solute from the soil (Selim, 1981). The reaction between S_2 and S_3 was considered to be of the kinetic first-order type, i.e.:

$$\frac{\partial S_3}{\partial t} = k_5 S_2 - k_6 S_3 \qquad (3\text{-}15)$$

where k_5 and k_6 (h^{-1}) are the reaction rate coefficients. If a consecutive reaction is included in the model, then eq. 3-13 must be modified to incorporate the reversible reaction between S_2 and S_3. As a result, the following equation

$$\rho \frac{\partial S_2}{\partial t} = k_3 \Theta C^n - \rho(k_4 + k_5)S_2 + \rho k_6 S_3 \qquad (3\text{-}16)$$

must be used in place of eq. 3-13. The above reactions are nonlinear in nature and represent initial-value problems that must be solved approximately using numerical methods. The initial conditions are that of a given initial solute concentration and assume no solute is retained at time zero, as is the case for kinetic batch experiments (see Amacher et al., 1988). Details of a numerical scheme (explicit–implicit finite difference) and documentation is given in Selim et al. (1990). In addition, the above retention mechanisms were incorporated into the classical convection–dispersion equation to predict solute retention as governed by the multireaction model during transport in soils (Selim et al., 1989). The multireaction and transport model, MRTM, is presented in Chapter 4.

SENSITIVITY ANALYSIS

We now consider the sensitivity of model predictions to the various model parameters. Figure 3-7 shows the effect of incorporating different parameters into the model. The effect is primarily in the change of the shape of the C vs. time curves. The magnitude of such a change is determined entirely by the magnitudes of the model parameters. In some cases, the effect may be large, while in others it is negligible. Amacher et al. (1988) showed that it was possible to fit a number of model variations to the same experimental data set so that the results were virtually indistinguishable. Thus, there are a number of combinations of model parameters and variations that will produce nearly the same results. Therefore, a unique solution of our model for prediction of a given data set should not be expected.

The effect of changing the reaction order associated with S_1 and S_2 on model simulation curves is shown in Figure 3-8. By decreasing the reaction order, the

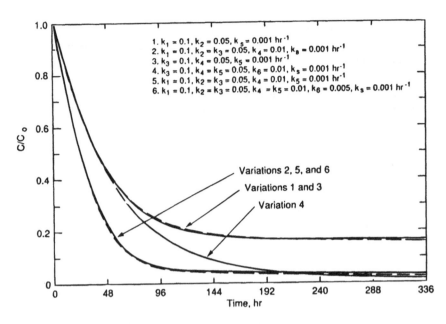

Figure 3-7 Effect of incorporating different parameters into MRM on model simulations. The parameters incorporated into each model variation and their values are shown in the figure. The reaction order for the nonlinear reversible reactions was 0.5.

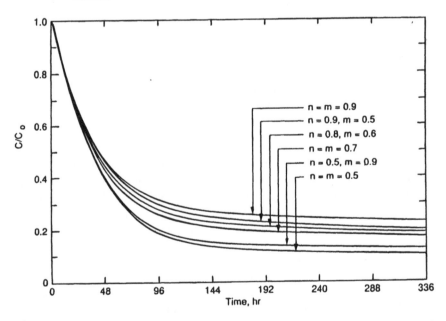

Figure 3-8 Effect of reaction order on MRM simulations. The reaction orders used in the model simulations are shown in the figure legend. The values of the rate coefficients are $k_1 = 0.1$, $k_2 = 0.05$, $k_3 = 0.01$, $k_4 = 0.005$, and $k_s = 0.001$.

rate of the reaction is increased. For three curves shown in Figure 3-8, the reaction orders for both nonlinear reversible reactions are the same ($n = m$), while for others the reaction orders are different.

The effect of the magnitude of the rate coefficients on the simulation curves for a given model variation is shown in Figure 3-9. As the magnitudes of the rate coefficients decrease, so do the reaction rates. By adjusting the rate coefficients, an almost infinite array of curve shapes is possible. Note that in Figures 3-7, 3-8, and 3-9, the curves are not appreciably affected by the model parameters at short reaction times. Only at longer times do the curves become separated.

APPLICATIONS

Application of MRM to batch kinetic data was carried out by Amacher et al. (1988, 1990) for retention of Cr(VI), Cd, and Hg by soils. Incorporation of MRM into the convective-dispersive transport equation and application of MRTM to transport of Cr(VI) in soils was done by Selim et al. (1989). Subsequently, Hinz et al. (1992) and Selim et al. (1992) applied MRM and MRTM to retention and transport of Zn and Cd in soils. These applications of MRTM to transport cases are considered in Chapter 4.

Estimation of Overall Reaction Order

To model retention and transport of heavy metals in soils using MRM and MRTM, an estimate of the overall reaction order (nonlinear parameters n and m

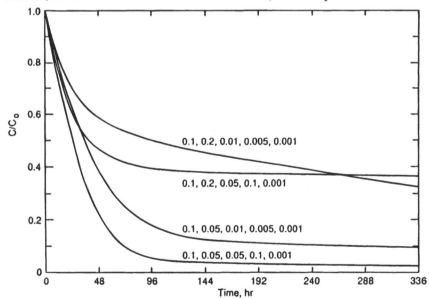

Figure 3-9 Effect of rate coefficients on model simulations. The values for k_1, k_2, k_3, k_4, and k_s are shown above each curve. The reaction order for the nonlinear reversible reactions was 0.5.

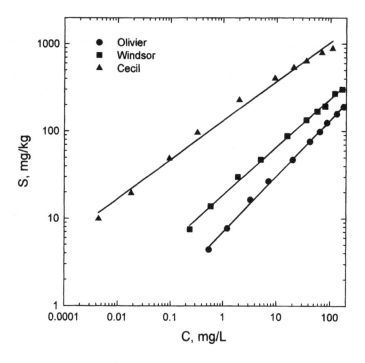

Figure 3-10 Freundlich sorption isotherms for Cr(VI) retention by Olivier, Windsor, and Cecil soils after 336 h of reaction.

in eqs. 3-12 and 3-13) is needed. For the case of Cr(VI) retention by three soils (Olivier, Windsor, and Cecil soils), values of n and m in eqs. 3-12 and 3-13 were estimated by fitting the Freundlich eq. 3-4 to the 336-h retention data (Figure 3-10 and Table 3-1). It was assumed that the reversible reactions were near equilibrium by this time. Little change in the slopes of the Freundlich sorption curves was observed after the first 24 h of reaction in these soils (Amacher et al., 1986), although a small continuing loss of Cr(VI) from solution was observed, prompting inclusion of the irreversible first-order reaction (eq. 3-14) in MRM. It was also assumed that n and m were numerically equivalent. Although, MRM allows for the possibility that n and m are different, there is as yet no unambiguous method for determining the reaction orders for most multiple concurrent reactions between a solute and soil sorbing surfaces. In modeling the batch kinetic data

TABLE 3-1 Freundlich Parameters K_f and b for Cr(VI) Retention by Olivier, Windsor, and Cecil Soils

Soil	r^2	K_f (cm^3 g^{-1})	b
Olivier	0.999	7 ± 1[a]	0.641 ± 0.009
Windsor	0.997	18.5 ± 1	0.55 ± 0.01
Cecil	0.991	132 ± 1	0.45 ± 0.01

[a] Parameter value ±95% confidence interval.

From Amacher, M.C. and H.M. Selim, 1994. *Ecol. Mod.* 74:205–230.

described here, the instantaneous adsorption equilibrium between S_e and C in Figure 3-6 (eq. 3-11) was not considered.

Variations of MRM

Figure 3-6 implies that numerous reaction combinations are possible, and thus a number of variations of MRM can be derived. Amacher et al. (1988) tested the ability of several variations of MRM to describe the time-dependent retention of Cr(VI) ($C = 1$ mg L^{-1}) by Windsor soil. This ranged from a model where all solute reactions (see Figure 3-6) were included to a model variation where only three phases (C, S_1 and S_{irr}) were considered. Amacher et al. (1988) showed that the MRM version—which accounted for two concurrent, nonlinear reversible reactions and one concurrent, first-order irreversible reaction— provided the best overall prediction of this data set. However, they showed that a number of model variations can produce simulations of the data that are statistically indistinguishable. It is not possible to distinguish among the possible reaction pathways by curve fitting alone. Independent experimental evidence is required. A similar conclusion was made by Skopp (1986). At present, there is no unambiguous experimental evidence to indicate which, if any, of the model variations is the "correct" one. Generally, the model that gives the best fit with the lowest parameter standard errors and has the least linear correlation among model parameters should be used. It is then desirable to obtain experimental evidence in support of a particular model choice. For example, it was not possible to determine whether the irreversible reaction is concurrent or consecutive, since both variations provided a similar overall description of the batch kinetic data. Consequently, due to its simplicity, the concurrent reaction version was utilized for further predictions of other data sets.

Description of Batch Kinetic Data with MRM

Examples of MRM predictions of time-dependent retention of Cr(VI) by Olivier, Windsor, and Cecil soils at several C_o values are shown in Figure 3-11, respectively (Amacher and Selim, 1994). A three-parameter (k_1, k_2, and k_s) version of MRM was satisfactory for describing the time-dependent retention of Cr(VI) by Olivier soil at low values of C_O. Because there was little change in Cr(VI) concentrations with time after about 24 h, a simple equilibrium adsorption model would be satisfactory for describing Cr(VI) retention at high values of C_o. For Windsor soil, a five-parameter (k_1, k_2, k_3, k_4, and k_s) version of MRM worked best at low values of C_o, whereas a three-parameter (k_1, k_2, and k_s) version was satisfactory at higher C_o values. As solute concentration increases, the mass of reactive solute relative to the masses of the different types of retention sites may increase to the point where the mass of solute greatly exceeds that of one type of retention site. At this point, the mass sorbed by that type of site is a negligible fraction of the total amount sorbed. Thus, only a single reversible, nonlinear retention reaction is needed in the model at high values of C_o. In contrast, a five-

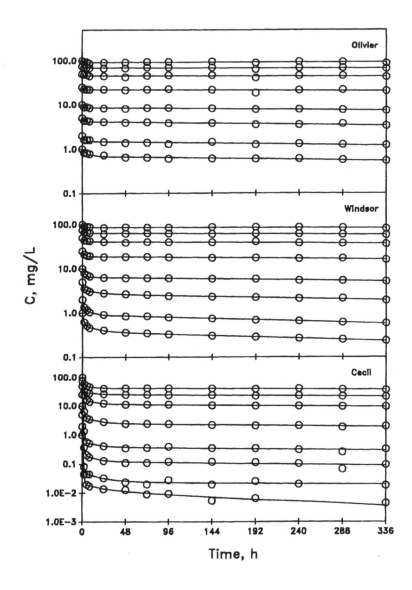

Figure 3-11   Time-dependent retention of Cr(VI) by Olivier, Windsor, and Cecil soils. Circles are data points and solid lines are MRM predictions for different initial concentration curves (C_o = 1, 2, 5, 10, 25, 50, 75, and 100 mg L^{-1}).

parameter model worked best at all values of C_o for Cecil soil, which retained a large amount of added Cr(VI).

Amacher et al. (1988, 1990) found that the magnitude of the rate coefficients was highly dependent on the initial (input) concentration C_o. Although the model is successful in describing kinetic data for a given C_o, the same rate coefficients cannot be used to describe data for substantially different initial concentrations. The systematic variation in the rate coefficients with C_o indicates that MRM is an incomplete description of the system and does not account for all the reactions and components. Rate coefficient constancy with variable reactant initial concentrations is required for the complete model. The model describes only the time-dependent changes in solute concentration and implicitly assumes that the concentrations of various types of reaction sites are not rate limiting, but are present in excess. Thus, MRM rate coefficients are pseudo rate coefficients and MRM is best considered as a representation of an apparent rate law, rather than a mechanistic rate law.

The reactions shown in Figure 3-6 should not be considered as elementary reactions that occur at the molecular level. The actual mechanism of retention is more complex and may include several sequential processes: mass transfer through the water phase (bulk diffusion), mass transfer through the hydrodynamic film surrounding the soil particles (film diffusion), mass transfer along the soil particle surface to the reaction sites (surface diffusion), or mass transfer through the soil particle to the reaction sites (particle diffusion) (Sparks, 1989). At the reaction sites, the actual chemical reaction may also proceed stepwise as a ligand exchange and may include formation of an outer-sphere surface complex first followed by elimination of water and formation of an inner-sphere complex (Sposito, 1984). All these steps are represented as a single process with an apparent rate coefficient for a given reaction site in Figure 3-6. Thus, the ratio of forward and reverse apparent rate coefficients does not give the equilibrium constant for the reaction at a given type of retention site because they are not rate coefficients for elementary reactions.

The diffusion processes are believed to be the rate-limiting steps in the above mechanism. Apparent rate coefficients that are film diffusion dependent will depend on mixing rate. In general, the apparent rate coefficient increases as mixing rate increases when film diffusion is rate limiting (Ogwada and Sparks, 1986). Burden (1989) obtained somewhat different MRM rate coefficients for Cr(VI) retention using a batch reactor with an overhead stirrer for the same soils than did Amacher et al. (1988), who used a reciprocating shaker. This evidence supports the interpretation that rate coefficients from batch experiments must be considered as apparent rate coefficients applicable to specific experimental conditions.

The systematic variation in MRM rate coefficients with C_o is also an indication that the concentration of reaction sites on the soil are also controlling the reaction rate. Further evidence for this can be seen by examining Figure 3-12, where the time-dependent relative concentrations (C/C_o) of the experimental data and MRM model predictions are shown for Cr(VI) retention by Olivier, Windsor, and Cecil soils, respectively. If the rate of retention depended solely on solute concentration (in a linear fashion), then the data points and lines should overlie one another. The changing shapes of the curves with C_o indicate a rate dependence on another reaction component, most likely the concentrations of reaction sites.

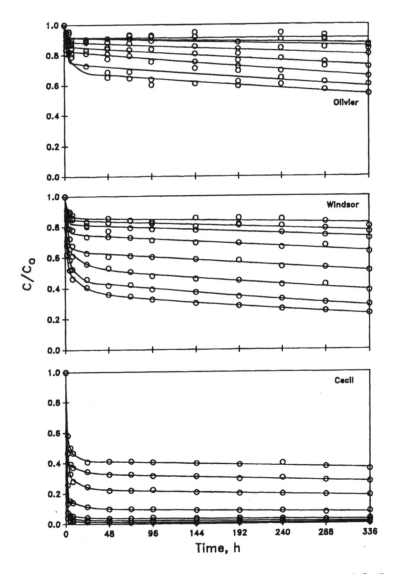

Figure 3-12 Time-dependent retention of Cr(VI) by Olivier, Windsor, and Cecil soils expressed in terms of relative concentrations. Circles are the data points and solid lines are MRM predictions for different initial concentration curves ($C_o = 1, 2, 5, 10, 25, 50, 75$, and 100 mg L^{-1}).

This would make the reaction second order overall, and this kinetic model is considered in Chapter 5.

The MRM model has proven to be versatile in describing the retention kinetics of several elements in soils (Amacher et al., 1988, 1990; Hinz et al., 1992; Selim et al., 1992). Examples of MRM descriptions of the retention kinetics of the group IIB elements, Zn, Cd, and Hg, in Windsor soil at several C_o are values shown in Figures 3-13, 3-14, and 3-15, respectively.

ONE SITE OR TWO SITES

The need for a two-site kinetic model in lieu of a one-site version is usually based upon the often better fit of a two-site model to batch kinetic or equilibrium sorption isotherm data than a one-site model. The actual chemical identity of these sites in soils is often unknown because of the absence of supporting experimental data. The model merely assumes that types of reaction sites can be distinguished on the basis of reaction rate, and does not require *a priori* knowledge of the microscopic structure of the sorbing surfaces and reaction pathways. Even in a system with a single sorbing surface (e.g., hydrous ferric oxide), sorption data are often interpreted on the basis of a two-site model (Dzombak and Morel, 1990). This finding is consistent with that of Kinniburgh (1986), who tested a number of isotherm equations for solute sorption on minerals and soils. He concluded that equations other than the single-site Langmuir and Freundlich models provided a better description of sorption phenomena. A lack of knowledge of the types of sites and reaction mechanisms need not hinder use of the models provided they are not extrapolated beyond the conditions of temperature, ionic strength, pH, and reactant concentrations used to obtain the batch kinetic data. Additional discussion of this point is given in Chapter 8.

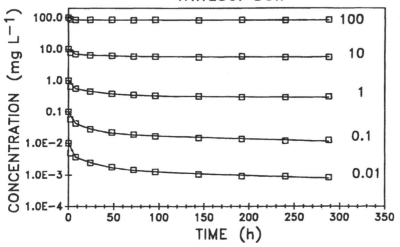

Figure 3-13   Time-dependent retention of Zn by Windsor soil. Squares are the data points and solid lines are MRM predictions for different initial concentration curves (C_o = 0.01, 0.1, 1, 10, and 100 mg L^{-1}).

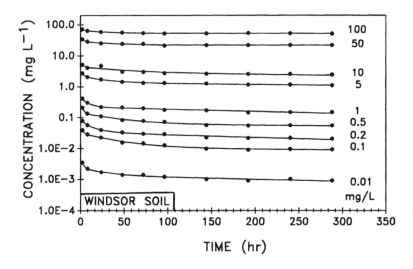

Figure 3-14 Time-dependent retention of Cd by Windsor soil. Circles are the data points and solid lines are MRM predictions for different initial concentration curves (C_o = 0.01, 0.1, 0.2, 0.5, 1, 5, 10, 50, and 100 mg L^{-1}).

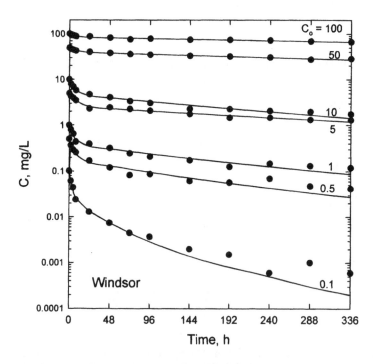

Figure 3-15 Time-dependent retention of Hg by Windsor soil. Circles are the data points and solid lines are MRM predictions for different initial concentration curves (C_o = 0.01, 0.1, 0.2, 0.5, 1, 5, 10, 50, and 100 mg L^{-1}).

The need for more than one reversible nonlinear reaction at low solute concentrations [e.g., Cr(VI) retention by Windsor and Cecil soils] is an indication that there may be more than one type of reaction site on the soil. Further evidence for this is provided by Figures 3-16, 3-17, and 3-18, where the cumulative percentages of Cr(VI), Cd, and Hg, respectively, released during dilution of the aqueous solution in contact with the soil are plotted against the amounts sorbed at the start of dilution (Amacher et al., 1988, 1990). The amounts of Cr(VI), Cd, and Hg subsequently released increased with the amounts initially sorbed. Thus, there would appear to be a range of different types of retention sites of differing energies or strengths of bonding of retained solute. At low levels of surface coverage, little of the retained solute was released back into solution when the solution in contact with the soil was diluted. As the amount retained increased, a proportional increase in the amount released was observed as lower bonding energy sites were filled and then emptied on dilution.

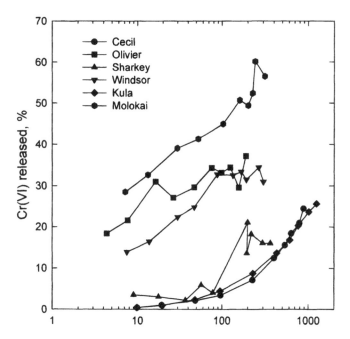

Figure 3-16 Cumulative release of Cr(VI) for different amounts of sorbed Cr(VI) on several soils.

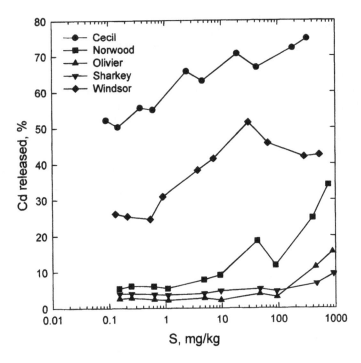

Figure 3-17 Cumulative release of Cd for different amounts of sorbed Cd on several soils.

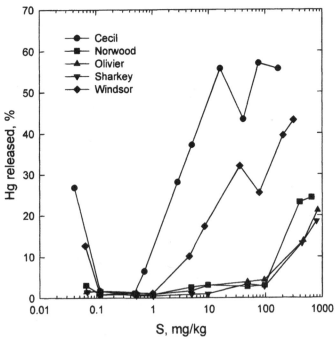

Figure 3-18 Cumulative release of Hg for different amounts of sorbed Hg on several soils.

4   TRANSPORT

To describe the general equation dealing with the transport of solutes present in the soil solution, a number of definitions must be given and the continuity or mass balance equation for the solute must be derived. One can assume that a heavy metal, or generally a solute species, may be present in a dissolved form in the soil water, i.e., the solution phase. The amount of a dissolved species is expressed in terms of concentration (mass per unit volume) in the solution phase. A solute species may also be retained by the soil matrix or be present in a precipitate or co-precipitate form.

For a given bulk volume within the soil, the total amount of solute χ(μg cm^{-3}) for a species i may be expressed as:

$$\chi_i = \Theta C_i + \rho S_i \qquad (4\text{-}1)$$

where S is the amount of solute retained by the soil (μg per gram soil), C is the solute concentration in solution (μg cm^{-3} or mg L^{-1}), Θ is the volumetric soil water content (cm^3 cm^{-3}) and ρ is the soil bulk density (g cm^{-3}).

CONTINUITY EQUATION

The continuity or mass balance equation for a solute species is a general representation of solute transport in the soil system and accounts for changes in solute concentration with time at any location in the soil. To derive the continuity equation, let us examine the transport of a solute species through a small volume element of a soil. For simplicity, we consider the volume element to be a small rectangular parallelepiped with dimensions Δx, Δy, and Δz, as shown in Figure 4-1. Assume that J_x is the flux or rate of movement of solute species i in the x-direction; that is, the mass of solute entering the face ABCD of the volume element per unit area and time. Therefore, the solute inflow rate, or total solute mass entering into ABCD per unit time, is

$$\textit{Solute inflow rate} = J_x \Delta y \Delta z \tag{4-2}$$

Similarly, if $J_{x+\Delta x}$ is the solute flux in the x direction for solute leaving the face EFGH, the total mass of solute leaving EFGH per unit time, i.e., the solute outflow rate, is

$$\textit{Solute outflow rate} = J_{x+\Delta x} \Delta y \Delta z \tag{4-3}$$

From elementary calculus, $J_{x+\Delta x}$ can be evaluated (approximately) from

$$J_{x+\Delta x} = J_x + \frac{\partial J_x}{\partial x} \Delta x \tag{4-4}$$

where $\partial J_x / \partial x$ is the rate of change of J_x in the x-direction.

From eq. 4-4, the net mass of solute flow (inflow minus outflow) per unit time in the volume element from solute movement in the x-direction is

$$\textit{Solute inflow rate} - \textit{Solute outflow rate} = (J_x - J_{x+\Delta x})\Delta y \Delta z$$

$$= -\frac{\partial J_x}{\partial x} \Delta x \Delta y \Delta z \tag{4-5}$$

Similarly, the net mass of solute flow per unit time from solute movement in the y-direction is

$$-\frac{\partial J_y}{\partial y} \Delta x \Delta y \Delta z \tag{4-6}$$

and from solute movement in the z-direction is

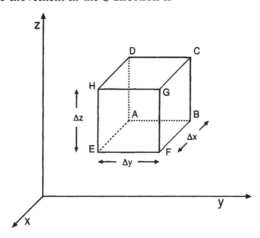

Figure 4-1 Rectangular volume element in the soil.

$$-\frac{\partial J_z}{\partial z} \Delta x \Delta y \Delta z \qquad (4\text{-}7)$$

Adding eqs. 4-5, 4-6, and 4-7 yields the net mass of solute (inflow–outflow) per unit time for the entire volume element as a result of solute movement in the x-, y-, and z-directions:

$$Net\ mass\ transport = -\left[\frac{\partial J_x}{\partial x} + \frac{\partial J_y}{\partial y} + \frac{\partial J_z}{\partial z}\right]\Delta x \Delta y \Delta z \qquad (4\text{-}8)$$

This net rate of solute flow represents the amount of mass of solute gained or lost within the volume element per unit time. This is often called the "rate of solute accumulation". Now we assume that the solute species considered, in our example, is of the nonreactive type; that is, the solute is not adsorbed or retained by the soil matrix. Therefore, we can further assume that for a nonreactive solute, S in eq. 4-1 is always zero and the solute is only present in the soil solution phase having a concentration C. Moreover, if Θ is the volumetric soil water content, i.e., volume of water per unit volume of bulk soil, then, $\Theta \Delta x \Delta y \Delta z$ is the total volume of water in the volume element shown in Figure 4-1. At any time, t, the total solute mass in the volume element is $\Theta C \Delta x \Delta y \Delta z$. Therefore, based on the principle of mass conservation, the rate of solute accumulation, i.e., rate of gain or loss $(\partial \Theta C/\partial t)\Delta x \Delta y \Delta z$, is equivalent to the net rate of mass flow (inflow–outflow). That is,

$$\frac{\partial \Theta C}{\partial t} \Delta x \Delta y \Delta z = -\left[\frac{\partial J_x}{\partial x} + \frac{\partial J_y}{\partial y} + \frac{\partial J_z}{\partial z}\right]\Delta x \Delta y \Delta z \qquad (4\text{-}9)$$

By dividing both sides of eq. 4-9 by the volume element $\Delta x \Delta y \Delta z$, we have

$$\frac{\partial \Theta C}{\partial t} = -\left[\frac{\partial J_x}{\partial x} + \frac{\partial J_y}{\partial y} + \frac{\partial J_z}{\partial z}\right] = -\operatorname{div} J \qquad (4\text{-}10)$$

which is called the *solute continuity equation for nonreactive solutes*. For the general case where the solute is of the reactive type, we can denote the extent of solute reaction in terms of the amount retained on the soil matrix S as described in eq. 4-1. Therefore, the rate of change of the total mass χ for the ith species with time may be represented by the following general solute continuity equation for rectangular coordinates as, (omitting the subscript i):

$$\frac{\partial \chi}{\partial t} = \frac{\partial \Theta C + \rho S}{\partial t} = -\operatorname{div} J \qquad (4\text{-}11)$$

The above equation is the general solute transport formulation dealing with the total amount of solute present in the soil system. Equation 4-11 does not include rates of production or removal of solutes from the soil, however. To achieve this, we introduce the therm Q to represent a sink or a source term that accounts for the rate of solute removal (or addition) irreversibly from a unit volume of a bulk soil (μg cm^{-3} h^{-1}). Incorporation of Q into eq. 4-11 yields:

$$\frac{\partial \Theta C + \rho S}{\partial t} = -\left[\frac{\partial J_x}{\partial x} + \frac{\partial J_y}{\partial y} + \frac{\partial J_z}{\partial z}\right] - Q \qquad (4\text{-}12)$$

This irreversible term Q can also be considered as a rate of volatilization or a root uptake term representing the rate of extraction (Q positive) of a solute from the bulk soil or the rate of exudation of a solute (Q negative). Moreover, in the following sections, we will restrict our analysis to one-dimensional flow in the z-direction where the flux J_z is dominant.

TRANSPORT MECHANISMS

It is commonly accepted that there are three mechanisms governing the transport of dissolved chemicals in porous media such as soils. Mass transport or convection refers to the passive transport of a dissolved chemical with water as the water moves through the soil. This mechanism is also referred to as *advection* or *piston flow*. If diffusion is ignored, the solute present in soil solution and the water move together at the same flow rate:

$$U_c = q_z C \qquad (4\text{-}13)$$

where q_z is the water flux density (cm^3 cm^{-2} h^{-1}) or simply the water flow velocity in the z-direction. This flow velocity q_z is referred to as *Darcy's flux* (cm h^{-1}).

Molecular diffusion is another transport mechanism that is due to the random thermal motion of molecules in solution. Diffusion is an active process regardless of whether or not there is net water flow in the soil. A good description of the diffusion process is Fick's law of diffusion, where the flux is proportional to the concentration gradient, and can be described (in the z-dimension) by:

$$U_d = -\Theta D_m \frac{\partial C}{\partial z} \qquad (4\text{-}14)$$

where D_m is the coefficient of molecular diffusion for the solute in the porous medium (cm^2 h^{-1}). This term D_m is also called the *apparent diffusion coefficient* and takes into account the effect of the solid phase of the porous medium on the diffusion. Van Schaick and Kemper (1966) related D_m to molecular diffusion of a solute species in pure water (D_o) according to:

$$D_o = D_m \tau \tag{4-15}$$

where τ is a tortuosity factor (dimensionless) that is likely to depend on the water content Θ. The factor τ takes a value less than 1 with a range of 0.3–0.7 for most soils. The transport of a solute in the tortuous flow path in porous medium is also accounted for by τ in eq. 4-15.

The third transport mechanism is due to dispersion or mixing of the solute in a porous medium. This is often referred to as *mechanical* or *hydrodynamic dispersion*, which includes all of the solute spreading mechanisms that are not attributed to diffusion. The mechanical or hydrodynamic dispersion phenomenon is due to the nonuniform flow velocity distribution during fluid flow in porous media (Ogata, 1970; Nielsen et al., 1986). Nonuniform velocity distribution through the soil pores is a result of variations in pore diameters along the flow path, fluctuation of the flow path due to tortuosity effects, and the variation in velocity from the center of a pore (maximum value) to zero at the solid surface interface (Poiseuille's law). The effect of dispersion is that of solute spreading, which is a tendency opposite to that of the so-called *piston flow*. When dispersion is neglected, the solute moves at the same identical velocity and a solute front arrives as one discontinuous jump. This behavior is called "piston flow" when a solute, if dispersion is ignored, moves in a porous medium or is displaced through the soil like a piston. Dispersion is a passive process in response to water flow. Therefore, dispersion is effective only during fluid flow, so that for a static water condition or when water flow is near zero, molecular diffusion is the dominant mixing process for solute transport in soils. Longitudinal dispersion (D_L) and transverse dispersion (D_T) coefficients are needed to describe the dispersion mechanism. Longitudinal dispersion refers to that in the direction of water flow, and D_T refers to dispersion in directions perpendicular or transverse to the direction of flow. Longitudinal dispersive transport can be described by an equation similar to eq. 4-14 for diffusion:

$$U_m = -\Theta D_L \frac{\partial C}{\partial z} \tag{4-16}$$

Separate studies investigated the effects of the soil water content Θ (Laryea et al., 1982; Smiles and Philip, 1978; Smiles et al., 1978) and the water flux q (Smiles and Gardiner, 1982; Koch and Flühler, 1993; De Smedt and Wierenga, 1984) on D_L. A linear relationship between D_L and v, where v is referred to as the pore water velocity (q/Θ), is commonly used,

$$D_L = D_o + \lambda v \tag{4-17}$$

where D_o is the molecular diffusion coefficient in water.

The term λ is a characteristic property of the porous media known as the *dispersivity* (cm). Dispersivity values λ vary from a few centimeters for uniformly packed (disturbed) laboratory soil columns to several meters for field-scale experiments. Large values of λ are also reported for well-aggregated soils.

In practice, an empirical parameter D rather than D_L is often introduced to simplify the flux equation (Boast, 1973). Moreover, because $D_o \ll D_L$, therefore D_L or $D = \lambda v$ or a more general formula (see Bear, 1972),

$$D = \lambda v^n \qquad (4\text{-}18)$$

is often used, where n is an empirical constant with a common range of 1.0 to 1.2 (Yasuda et al., 1994; Montero et al., 1994; Jaynes et al., 1988). Therefore, D vs. v may not be strictly linear and the dispersivity is not velocity independent (Gerritse and Singh, 1988).

Several modifications of D vs. v are found in the literature by introducing the tortuosity coefficient τ of eq. 4-15, such as $D = D_o\tau + \lambda v$ (Brusseau, 1993; Jensen, 1983) and $D = (D_o + \lambda v)\tau$ (Rao et al., 1976). More general regression formulae include $D = a + bv$ (De Smedt and Wierenga, 1984, Wierenga, 1977) and $D = a + bv^2$ (Koch and Flühler, 1993), where a and b are constants, with a much greater than D_o. Another regression formula, $D = D_m(\Gamma + aP_e^b)$, is also used (Bond and Smiles, 1983; Rose, 1977; Bond, 1986), where Γ is a constant and P_e is the particle Peclet number defined as $P_e = vl/D_o$ (l is a characteristic length of the porous material).

Besides water flow velocity, D or D_L is also affected by the degree of water saturation. D_L is usually much larger under unsaturated soils (De Smedt and Wierenga, 1984; Laryea et al., 1982). In a horizontal infiltration study with a silty clay loam, Laryea et al. (1982) estimated increased D values as the soil water content increased. Moreover, D depends on porosity and pore size distribution. Koch and Flühler (1993) found that D values were much larger in porous beads than in spherical solid beads.

Combining eqs. 4-13, 4-14, and 4-16 yields the following simplified solute flux expression in the z-direction:

$$J_z = -\Theta D \frac{\partial C}{\partial z} + q_z C \qquad (4\text{-}19)$$

which incorporates the effects of mass flow or convection as well as diffusion and mechanical dispersion. Incorporation of flux eq. 4-19 into the conservation of mass eq. 4-13 yields the following generalized form for solute transport in soils in one-dimension:

$$\frac{\partial \Theta C}{\partial t} + \rho \frac{\partial S}{\partial t} = \frac{\partial}{\partial z}\left[\Theta D \frac{\partial C}{\partial z}\right] - \frac{\partial q_z C}{\partial z} - Q \qquad (4\text{-}20)$$

The above equation is commonly known as the convective-dispersive equation

(CDE) for solute transport, and is valid for soils under transient and unsaturated soil-water flow conditions.

In order to describe the fate of solutes in unsaturated soil profiles under transient flow conditions, the Richards equation for water flow (in one dimension) must also be considered (Nielsen et al., 1986):

$$\frac{\partial \Theta}{\partial t} = \frac{\partial}{\partial z}\left[K(h)\,\frac{\partial h}{\partial z} \right] - \frac{\partial K(h)}{\partial z} - A(z,t) \qquad (4\text{-}21)$$

where h is the soil-water pressure head (cm) and $K(h)$ is the soil hydraulic conductivity (cm h^{-1}). The above equation is known as the h-form of the water flow equation. Knowledge of the functional relation of K vs. pressure head h and the moisture characteristic relation (Θ vs. h) are prerequisites for solving the Richards flow equation. In addition, eq. 4-21 includes a root uptake term, $A(z,t)$, for water extraction as a function of depth and time (cm^3 cm^{-3} h^{-1}). This term is analogous to the irreversible source/sink term Q for solutes in eq. 4-20. Upon solution of eq. 4-21, subject to the appropriate initial and boundary conditions, one can obtain the water content Θ and Darcy flux q for any depth (z) and time (t). Both $\Theta(z,t)$ and $q(z,t)$ are needed in order to obtain a solution for solute transport in the convection–dispersion eq. 4-20.

STEADY-STATE FLOW

For conditions where steady water flow is dominant, q and Θ are constants over space and time; that is, for uniform Θ in the soil, we have the simplified form of the convection–dispersion equation as:

$$\frac{\partial C}{\partial t} + \frac{\rho}{\Theta}\,\frac{\partial S}{\partial t} = D\,\frac{\partial^2 C}{\partial z^2} - v\,\frac{\partial C}{\partial z} - \frac{Q}{\Theta} \qquad (4\text{-}22)$$

where v is the pore-water velocity (q_z/Θ).

We further assume that the amount retained by the soil (S) is related to solute concentration in solution (C) by a linear or linearized equilibrium isotherm of the form:

$$S = K_d C \qquad (4\text{-}23)$$

where K_d is an empirical distribution coefficient (cm^3 g^{-1}).

This simple assumption of linear adsorption is generally valid for solutes of low retention or affinity to the soil and at low concentrations. Moreover, we assumed that the solute is thus not subject to degradation, decay, or production and thus the irreversible term Q can be ignored. Equation 4-22 reduces to:

$$R\frac{\partial C}{\partial t} = D\frac{\partial^2 C}{\partial z^2} - v\frac{\partial C}{\partial z} \tag{4-24}$$

where R is the retardation factor:

$$R = 1 + \frac{\rho}{\theta} K_d \tag{4-25}$$

If there is no solute retention by the soil, K_d becomes zero and R becomes one. Such an assumption is often made for anionic and neutral tracers such as chloride, bromide, tritium, among others. In some cases, R may become less than one, indicating that only a fraction of the soil solution phase participates in the transport process. This may be the case when the solute is subject to significant anion exclusion or when relatively immobile water regions are present (e.g., inside dense aggregates) that do not contribute to convective transport. van Genuchten and Wierenga (1986) suggested that, in case of anion exclusion, $(1 - R)$ may be viewed as the relative anion exclusion volume.

BOUNDARY AND INITIAL CONDITIONS

Solutions of the above convection–dispersion eq. (4-7) or (4-9) yield the concentration distribution of the amount of solute in soil solution (C) and that retained by the soil matrix (S) with time and space in soil (z,t). In order to arrive at such a solution, the initial and boundary conditions that accurately describe experimental conditions must be specified. Several boundary conditions are identified with the problem of solute transport in porous media. First-type (Dirichlet) boundary conditions for a solute pulse input may be described as:

$$C = C_o, \quad z = 0, \quad t < t_p \tag{4-26}$$

$$C = 0, \quad z = 0, \quad t \geq t_p \tag{4-27}$$

where $C_o(\mu g\ cm^{-3})$ is the concentration of the solute species in the input pulse.

The input pulse application is for a duration $t_p(h)$, which is then followed by solution that is free of solute. These boundary conditions describe a tracer solution applied at a specified rate from a perfectly mixed inlet reservoir to the surface of a finite or semi-infinite soil profile. These Dirichlet boundary conditions were used by Lapidus and Amundson (1952) and Clearly and Adrian (1973) and assume that the concentration itself can be specified at the inlet boundary. This situation is not usually possible in practice.

Third-type boundary conditions are commonly used, and account for advection plus dispersion across the interface of solute at concentrations C_o. For a continuous input at the soil surface, we have

$$vC_o = -D\frac{\partial C}{\partial z} + vC, \quad z = 0, \quad t > 0 \tag{4-28}$$

and for a third type pulse-input, we have

$$vC_o = -D\frac{\partial C}{\partial z} + vC, \quad z = 0, \quad t < t_p \tag{4-29}$$

$$0 = -D\frac{\partial C}{\partial z} + vC, \quad z = 0, \quad t \geq t_p \tag{4-30}$$

These conditions are for a solute or a tracer solution that is applied at a specified rate from a perfectly mixed inlet reservoir to the soil surface. Continuity of the solute flux across the inlet boundary leads directly to the above third-type boundary conditions. Third-type boundary conditions were used by Brenner (1962) and Lindstrom et al. (1967), among others.

Proper formulation of the exit boundary condition for displacement through finite laboratory columns and soils at the field scale are needed. The boundary condition at some depth L in the soil profile is often expressed as (Danckwerts, 1953):

$$\frac{\partial C}{\partial z} = 0, \quad z = L, \quad t \geq 0 \tag{4-31}$$

This second-type boundary condition is used to deal with solute effluent from soils having finite depths such as laboratory miscible displacement columns and describes a zero concentration gradient at $z = L$.

It is often convenient to solve the dispersion–convection equation where a semi-infinite rather than a finite length (L) of the soil is assumed. Under such circumstances, the appropriate condition for a semi-infinite medium is needed. Specifically, for semi-infinite systems in the field we need a boundary condition that specifies solute behavior at large depth ($z \to \infty$). Such a boundary condition may be expressed as $C =$ constant (commonly zero) as $z \to \infty$. However, appropriate formulation of this boundary is

$$\frac{\partial C}{\partial z} = 0, \quad z \to \infty, \quad t \geq 0 \tag{4-32}$$

which is identical to that for a finite soil length (L). Kreft and Zuber (1978) and van Genuchten and Wierenga (1986) presented a discussion of the various types of boundary conditions for solute transport problems.

EXACT SOLUTIONS

Analytical or exact solutions to the convection–dispersion (CD) equation 4-24 (or eq. 4-24 subject to the appropriate boundary and initial conditions) are available for a limited number of situations whereas the majority of the solute transport problems must be solved using numerical approximation methods. In general, whenever the form of the retention reaction is linear, an exact or closed-form solution is obtainable. A number of closed-form solutions are available in the literature and are compiled by van Genuchten and Alves (1982). Since several boundary conditions are commonly used, we limit our discussion to three exact solutions to the CD equation (4-24) widely cited in the literature. All three solutions have the form:

$$C(z,t) = \begin{cases} C_i + (C_o - C_i)A(z,t) & 0 < t < t_p \\ C_i + (C_o - C_i)[A(z,t) - A(t,t - t_p)] & t > t_p \end{cases} \quad (4\text{-}33)$$

where we assume a simple initial condition ($C = C_i$) representing uniform solute concentration distribution in the soil at $t = 0$. The form of the solution in eq. 4-33 is applicable for conditions representing continuous solute application or a pulse-type input having duration t_p. For the three exact solutions, the appropriate expressions for $A(z,t)$ are given below.

Brenner Solution

Brenner (1962) considered the case of a finite soil column with the more precise third-type boundary condition at $z = 0$ that accounts for dispersion and advection across the upper surface (eqs. 4-29 and 4-30). By defining the Peclet number as:

$$P = \frac{vL}{D} \quad (4\text{-}34)$$

Brenner's solution can be expressed as:

$$A(z,t) = 1 - \sum_{m=1}^{\infty} \frac{2P\beta_m \left[\beta_m \cos\left(\frac{\beta_m z}{L}\right) + \frac{P}{2} \sin\left(\frac{\beta_m z}{L}\right) \right] \exp\left[\frac{zP}{2L} - \frac{Pvt}{4LR} - \frac{\beta_m^2 vt}{PLR} \right]}{\left[\beta_m^2 + \frac{P^2}{4} + P \right]\left[\beta_m^2 + \frac{P^2}{4} \right]}$$

$$(4.35)$$

where the eigenvalues β_m are the positive roots of:

$$\beta_m \cot(\beta_m) - \frac{\beta_m^2}{P} + \frac{P}{4} = 0 \quad (4\text{-}36)$$

Brenner's solution describes volume-averaged concentrations within the column.

Because of the zero concentration gradient at $z = L$, this solution also defines a flux concentration at the lower boundary. Hence, Brenner's solution correctly interprets effluent concentrations as representing flux-averaged concentrations. Another feature of Brenner's solution is that the mass balance requirement for a finite column is met. That is, the amount of solute that is entering the column minus the amount that is leaving that column equals that which is stored in the column. Brenner's solution, which is a series, converges only for relatively small values of P. In fact, Selim and Mansell (1976) found that Brenner's solution requires as many as 100 terms to obtain convergence for $P > 20$. Therefore, approximate solutions are recommended for large P values.

Lindstrom Solution

Lindstrom et al. (1967) considered the case for a semi-infinite medium with a third-type boundary at the soil surface. Specifically, the boundary conditions given by eqs. 4-29, 4-30, and 4-32 were used. This case is similar to that considered by Brenner (1962) except for a semi-infinite rather than finite column lengths. Lindstrom's solution can be expressed as:

$$A(z,t) = \frac{1}{2} \, erfc\left[\frac{Rz - vt}{(4DRt)^{1/2}}\right] + \left[\frac{v^2 t}{\pi DR}\right]^{1/2} \exp\left[-\frac{(Rz - vt)^2}{4DRt}\right]$$

$$-\frac{1}{2}\left[1 + \frac{vz}{D} + \frac{v^2 t}{DR}\right] \exp\left(\frac{vz}{D}\right) erfc\left[\frac{Rz + vt}{(4DRt)^{1/2}}\right] \quad (4\text{-}37)$$

This solution does not suffer from convergence problems as does that of Brenner's. In the meantime, it provides accurate mass balance and describes volume-averaged concentrations in the column.

Cleary and Adrian Solution

Cleary and Adrian (1973) considered a similar case to that of Brenner (1962) except for a first-type boundary condition at the inlet ($z = 0$). Their solution is for finite column lengths having boundary conditions eqs. 4-26, 4-27, and 4-31 and may be expressed as:

$$A(z,t) = 1 - \sum_{m=1}^{\infty} \frac{2\beta_m \, \sin\left(\frac{\beta_m z}{L}\right) \exp\left[\frac{zP}{2L} - \frac{Pvt}{4LR} - \frac{\beta_m^2 vt}{PLR}\right]}{\left[\beta_m^2 + \frac{P^2}{4} + \frac{P}{2}\right]} \quad (4\text{-}38)$$

where the eigenvalues β_m are the positive roots of:

$$\beta_m \cot(\beta_m) + \frac{P}{2} = 0 \qquad (4\text{-}39)$$

This solution suffers from the same convergence problems as Brenner's, which converges only for relatively small values of P. Moreover, Cleary and Adrian's solution fails the mass-balance requirement and also violates mass balance for the effluent curve. van Genuchten and Wierenga (1986) recommended that this solution not be applied to solute transport column experiments.

The significance of using the precise boundary conditions is illustrated by comparing the Cleary and Adrian (1973) solution with the concentration profiles calculated using the mathematical solutions of Brenner (1962) and Lindstrom et al. (1967). Unlike the boundary condition used by the other two solutions, Cleary and Adrian (1973) assumed a first-type boundary condition ($C = C_o$ at $z = 0$), which has been used by several investigators (Gupta and Greenkorn, 1973; Kirda et al., 1973; Lai and Jurinak, 1972; Warrick et al., 1971). The solution by Lindstrom et al. (1967) was developed for a semi-infinite soil column but was later applied to finite soil columns by Davidson et al. (1968) and Davidson and Chang (1972), among others.

The above solutions were used to calculate distributions of solute concentration in a 30-cm soil column (Figure 4-2) for selected times during continuous application of a solute solution (see Selim and Mansell, 1976). These concentration profiles were obtained for pore-water velocities v of 0.5, 1.5, and 3.0 cm h^{-1}. Parameters used were those of Lai and Jurinak (1972), where $D = 1.5$ cm^2 h^{-1}, $\rho = 1.30$ g cm^{-3}, $\Theta = 0.45$ cm^3 cm^{-3}, $K_d = 2.5$ cm^3 g^{-1}, and $L = 30$ cm with an R value of 8.22.

As was expected, the solution of Cleary and Adrian provided higher concentrations throughout the soil column at all times and for all three pore velocities (Figure 4-2) in comparison with results obtained by using the other two solutions (see Selim and Mansell, 1976). With decreasing pore water velocity (v), the magnitude of deviation among the three solutions increased. The higher concentrations are attributed to the assumption that $C = C_o$ at $z = 0$ for all times. However, the concentration profiles obtained using the solution of Lindstrom et al. (1967) are essentially identical to results obtained using the Brenner (1962) solution except in the vicinity of the exit end of the soil column ($z = L = 30$ cm) at large times. Deviation of the Lindstrom et al. (1967) solution in the vicinity of $z = L$ is clearly due to forcing a semi-infinite boundary condition to describe solute transport for the finite soil columns presented here.

Solute breakthrough curves corresponding to pore water velocities of 0.5, 1.5, and 3.0 cm h^{-1} for the same soil parameters as in Figure 4-2 are shown in Figure 4-3 (see Selim and Mansell, 1976). Breakthrough curves are commonly used in miscible displacement studies and represent relative solute concentration C/C_o at $z = L$ vs. relative volume of accumulated effluent V/V_o, where V_o is the pore water volume in the soil column ($V_o = \Theta L$). For all three pore water velocities, the breakthrough curves obtained using the Brenner solution occur between those obtained by using solutions of Cleary and Adrian and Lindstrom

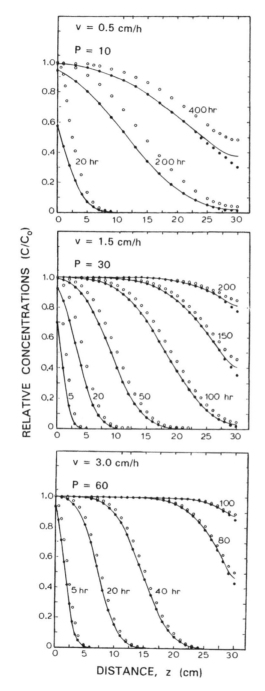

Figure 4-2 Relative concentration (C/C_o) vs. distance z (cm) within a soil at various times for $P = 10$, 20, and 60 ($v = 0.5$, 1.5, and 3.0 cm, respectively). Solid curves were obtained using the Brenner solution (eq. 4-35), open circles using the Clearly and Adrian solution (eq. 4-38), and closed circles by the Lindstrom et al. solution (eq. 4-37).

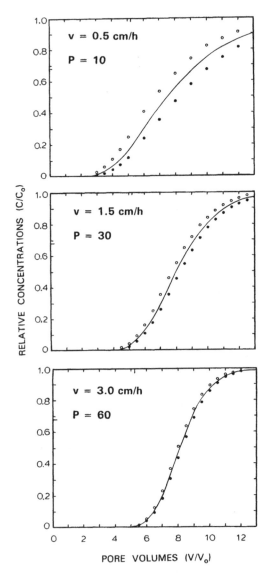

Figure 4-3 Relative concentration (C/C_o) vs. pore volumes at $L = 30$ cm within a soil for $P = 10, 20,$ and 60 ($v = 0.5, 1.5,$ and 3.0 cm, respectively). Solid curves were obtained using the Brenner solution (eq. 4-35), open circles using the Clearly and Adrian solution (eq. 4-38), and closed circles by the Lindstrom et al. solution (eq. 4-37).

et al. As is indicated in Figure 4-3, deviation between the Brenner solution and the other two solutions decreases as the pore water velocity or Peclet number P increases. For $P = 10$, the deviation was about 8%; whereas for $P = 60$, the deviation was only 2%. Calculated breakthrough curves for several values of the column Peclet number P are shown in Figure 4-4. As P decreases, the extent of

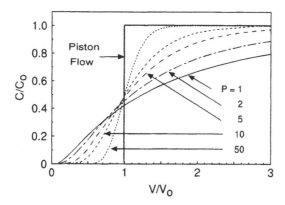

Figure 4-4   Relative concentration (C/C_o) vs. pore volume (V/V_o) based on the Lindstrom et al. solution (eq. 4-37) for $R = 1$ and various values of P.

spreading increases due to increased apparent dispersion coefficient D or decrease in v. Moreover, as $P \to \infty$ ($D \to 0$ or $v \to \infty$) the breakthrough resembles that for a step-function at one pore volume ($V/V_o = 1$) and represents condition of piston-type flow or convection only.

ESTIMATION OF D

In a number of field and laboratory miscible displacement studies, the main purpose of a tracer application is to estimate the apparent dispersion coefficient, D. A commonly used technique for estimating D is to describe tracer breakthrough results where tritium, chloride-36, bromide, or other tracers are used. It is common to use one of the above exact solutions or an approximate (numerical) solution of the CD equation. In addition, a least-squares optimization scheme, or curve-fitting method, is often used to obtain best-fit estimates for D. One commonly used curve-fitting method is the maximum neighborhood method of Marquardt (1963), which is based on an optimum interpolation between the Taylor series method and the method of steepest descent (Daniel and Wood, 1973) and is documented in a computer algorithm by van Genuchten (1981).

The goodness of fit of tracer BTCs is usually unacceptable when D is the only fitting parameter. Thus, two parameters are fitted (usually D along with the retardation factor R) in order to improve the goodness-of-fit of tracer BTCs (van Genuchten 1981). Other commonly fitted parameters include pulse duration t_p and the flow velocity v for solute retention (Jaynes et al., 1988; Andreini and Steenhuis, 1990). However, since v can be experimentally measured under steady-state flow, it may not be appropriate to fit v to achieve improved fitting of BTCs. The best-fit velocity v is often different from that experimentally measured. Estimates for R values for tritium and chloride-36 tracers are often close to unity for most soils. R values greater than unity indicate sorption or simply retardation, whereas $R < 1$ may indicate ion exclusion or negative sorption. Similar values for R for tritium and ^{36}Cl were reported by Nkedi-Kizza et al. (1983), van Genuchten and Wierenga (1986), and Selim et al. (1987). Table 4-1 provides estimates for D values as obtained from tracer breakthrough results for several soils. Selected examples of measured and best-fit prediction of tritium break-through results are shown in Figures 4-5 and 4-6 for two reference clays (bentonite and kaolinite) and a Sharkey clay soil material (Figure 4-7) (Gaston and Selim, 1990b, 1991). Ma and Selim (1994) proposed the use of an effective path length L_e or a tortuosity parameter $\tau (L_e/L)$, where L_e was obtained based on mean residence time measurements. They tested the validity of fitting solute transport length (L_e) or tortuosity (τ) using the CD equation and concluded that L_e (or τ) can serve as a second parameter in the fitting of tracer results in addition to the dispersion coefficient (D). The use of an effective path length is physically more meaningful than the use of R as a second fitting parameter. Furthermore, independent verification of R values for tracer solutes when R is significantly different from one is not possible.

TABLE 4-1 Values of the Dispersion Coefficients (D) for Selected Soils and Minerals from Tritium Pulse Miscible Displacement (Soil Column) Experiments

Soils	Column length (cm)	Pore-water velocity (v) (cm h^{-1})	Dispersion coefficient (D) (cm^2 h^{-1})
Sharkey Ap	15	4.00	4.96
Sharkey (4-6 mm)	10	1.75	9.02
Sharkey (4-6 mm)	15	1.85	7.66
Eustis (<2 mm)	15	2.66	2.02
Cecil (<2 mm)	15	1.07	2.39
Cecil (<2 mm)	15	2.23	8.32
Cecil (<2 mm)	15	5.21	20.8
Cecil (0.5–1.0 mm)	15	2.05	3.71
Mahan (<2 mm)	15	2.02	9.79
Mahan (<2 mm)	15	3.82	15.8
Mahan (<2 mm)	15	5.29	23.0
Dothan Ap (<2 mm)	15	2.74	11.0
Dothan Bt (<2 mm)	15	2.32	11.0
Olivier (Ap)	10	2.28	1.01
Windsor (Ap)	10	1.13	0.27
Windsor (Ap)	10	3.29	3.77
Yolo (Ap)	10	1.16	0.17
Acid wash sand	15	2.92	1.10
Glass beads + sand[a] (1:1 by weight)	15	3.08	0.73
Kaolinite + sand (1:1 by weight)	10	0.52	0.40
Bentonite + sand (1:9 by weight)	10	0.63	0.56

[a] Particle size distributions (0.25–0.50 mm, 0.50–1.0 mm, and 1–2 mm) were 0, 97.23%, and 2.13% for glass beads, and 79.96%, 19.23%, and 0 for acid washed sand, respectively.

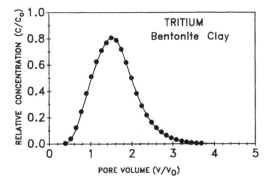

Figure 4-5 Tritium breakthrough results (C/C_o vs. V/V_o) for a bentonite:sand column (with 1:9 clay:sand mixture). Solid curve is fitted BTC curve.

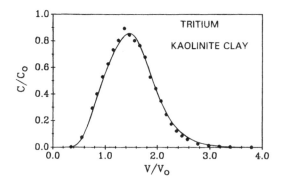

Figure 4-6 Tritium breakthrough results (C/C_o vs. V/V_o) for a kaolinite:sand column (1:1 mixture). Solid curve is fitted BTC curve.

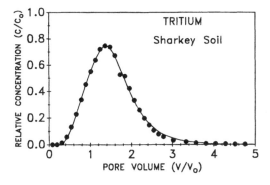

Figure 4-7 Tritium breakthrough results (C/C_o vs. V/V_o) for a Sharkey clay soil. Solid curve is fitted BTC curve.

OTHER EXACT SOLUTIONS

In addition to the exact solutions presented above (eqs. 4-35 to 4-39) for the CD equation presented, several other exact solutions are also available. Specifically, a number of solutions are available for the CD equation, having the form:

$$R \frac{\partial C}{\partial t} = D \frac{\partial^2 C}{\partial z^2} - v \frac{\partial C}{\partial z} - a_1 C - a_2 \qquad (4\text{-}40)$$

which includes a linear (reversible) retention term as described by the retardation factor R (of eq. 4-25). In addition, eq. 4-40 includes a first-order irreversible reaction term, with a_1 the associated rate coefficient (h^{-1}). In addition, it also includes a zero-order sink source/source term having a constant rate of loss a_2 (or gain for negative a_2). Subject to different sets of boundary conditions (eqs. 4-26 to 4-32) for soil columns of finite or semi-infinite lengths, a number of exact solutions to eq. 4-40 are available in the literature (Carslaw and Jaeger, 1959; Ozicik, 1968; Selim and Mansell, 1976; van Genuchten and Alves, 1982). However, most retention mechanisms are nonlinear and time-dependent in nature, and analytical solutions are not available. As a result, a number of numerical models using finite-difference or finite-element approximations have been utilized to solve nonlinear retention problems of multiple reactions and multicomponent solute transport for one- and two-dimensional geometries.

NONLINEAR RETENTION AND TRANSPORT

Several examples of heavy metal retention by different soils were illustrated in Chapter 3. A common feature is their nonlinear behavior, as well as the time-dependent or kinetic-retention characteristics. Consequently, we present a general nonlinear approach that deals with time-dependent and nonlinear retention of heavy metals during transport in soils. This approach represents an extension of the multireaction model (MRM) described in Chapter 3. The classical convective-dispersive transport used is given by eq. 4-22. In a similar fashion to the multireaction model discussed by Amacher et al. (1988) and Selim et al. (1989), we consider the $(\partial S/\partial t)$ term of eq. 4-22 to describe fully reversible processes between heavy metals in the solution and that on the solid phase. Specifically, we consider the reversible retention to be of the multireaction (multi-site) equilibrium-kinetic type where S is composed of four phases:

$$S = S_e + S_1 + S_2 + S_3 \qquad (4\text{-}41)$$

Here, we assume S_e to be the amount of metal solute (mg per kg soil) that is sorbed reversibly and is in equilibrium with that in soil solution phase (C) at all times. The governing equilibrium retention/release mechanism is the nonlinear Freundlich type:

$$S_e = K_f C^b \tag{4-42}$$

where K_f is the associated distribution coefficient ($cm^3 \, kg^{-1}$) and b is a dimensionless Freundlich parameter ($b < 1$).

Other equilibrium type retention mechanisms are given in Chapter 3. This includes linear ($b = 1$), Langmuir, two-site Langmuir, and others.

The retention/release reactions associated with S_1, S_2, and S_3 are concurrent or consecutive type kinetic reactions (see Figure 3-1, Chapter 3). Specifically, the S_1 and S_2 phases were considered to be in direct contact with C and reversible rate coefficients of the (nonlinear) kinetic type govern their reactions:

$$\frac{\partial S_1}{\partial t} = k_1 \frac{\Theta}{\rho} C^n - k_2 S_1 \tag{4-43}$$

$$\frac{\partial S_2}{\partial t} = k_3 \frac{\Theta}{\rho} C^m - (k_4 + k_5) S_2 + k_6 S_3 \tag{4-44}$$

$$\frac{\partial S_3}{\partial t} = k_5 S_2 - k_6 S_3 \tag{4-45}$$

where k_1 and k_2 are the forward and backward rate coefficients (h^{-1}), respectively, and n is the reaction order associated with S_1. Similarly, k_3 and k_4 are the rate coefficients and m is the reaction order associated with S_2; k_5 and k_6 are the rate coefficients associated with S_3. In the absence of the consecutive reaction between S_2 and S_3, that is, if $S_3 = 0$ at all times ($k_5 = k_6 = 0$), eq. 4-44 reduces to:

$$\frac{\partial S_2}{\partial t} = k_3 \frac{\Theta}{\rho} C^m - k_4 S_2 \tag{4-46}$$

Thus, eq. 4-46 for S_2 resembles that for S_1 except for the magnitude of the associated parameters k_3, k_4 and m.

The sorbed phases (S_e, S_1, S_2, and S_3) may be regarded as the amounts sorbed on surfaces of soil particles and chemically bound to Al and Fe oxide surfaces or other types of surfaces, although it is not necessary to have a priori knowledge of the exact retention mechanisms for these reactions to be applicable. These phases may be characterized by their kinetic sorption and release behavior to the soil solution and thus are susceptible to leaching in the soil. In addition, the primary differences between these two phases not only lie in the difference in their kinetic behavior but also on the degree of nonlinearity, as indicated by the parameters n and m.

The sink/source term Q of eq. 4-22 is commonly used to account for irreversible reactions such as precipitation/dissolution, mineralization, and immobilization, among others. We expressed the sink term as a first-order kinetic process:

$$Q = \rho \frac{\partial S_3}{\partial t} = k_s \Theta C \qquad (4\text{-}47)$$

where k_s is the associated rate coefficient (h^{-1}).

The sink term Q was expressed in terms of a first-order irreversible reaction for reductive sorption or precipitation or internal diffusion, as described by Amacher et al. (1986, 1988). Equation 4-43 is similar to that for diffusion controlled precipitation reaction if one assumes that the equilibrium concentration for precipitation is negligible (Stumm and Morgan, 1981).

The above formulation is a general one (eqs. 4-43 to 4-47) and includes equilibrium and kinetic retention processes. In fact, the multireaction model developed by Amacher et al. (1988), on which this transport and retention model is based, is a fully kinetic one where local equilibrium with the solution phase was not implicitly considered. In addition, the two-site model proposed by Selim et al. (1976) and Cameron and Klute (1977) was the earliest to include a linear or nonlinear kinetic reaction and an equilibrium reaction that were incorporated with the convection–dispersion transport eq. 4-22.

In order to solve the transport and retention/release equations associated with the above multireaction and transport model (MRTM), the appropriate initial and boundary conditions must be specified. For the sensitivity and application cases presented in following sections, we restrict our discussion to steady-state water flow conditions in a homogeneous soil having a uniform moisture distribution. Therefore, the water flux q and volumetric soil water content (Θ) are considered time invariant. It is also assumed that a solute solution of known concentration (C_o) is applied at the soil surface for a given duration t_p and was thereafter followed by a solute free solution. These conditions are correctly described by eqs. 4-29 through 4-32. Moreover, the initial conditions used were

$$C = C_i, \qquad t = 0, \qquad 0 < z < L \qquad (4\text{-}48)$$

$$S_e = S_1 = S_2 = S_3 = 0, \qquad t = 0, \qquad 0 < z \le L \qquad (4\text{-}49)$$

These conditions represent a soil profile of length, L, (cm) having uniform initial concentration C_i in the solution and devoid of sorbed phases along the soil profile at time zero. However, this model is not restricted to uniform conditions. Rather, nonuniform initial distributions of C, S_e, S_1, S_2, and S_3 can be used.

NUMERICAL SOLUTION

The convection–dispersion solute transport eq. 4-24, along with the retention eqs. 4-42 to 4-47 of the nonlinear multireaction approach, can be solved using numerical approximations since closed-form solutions are not available. Commonly used numerical methods are the finite difference explicit-implicit methods (Remson al., 1971; Pinder and Gray, 1977). Finite difference solutions provide distributions of solution (C) and sorbed phase concentrations (S_e, S_1, S_2, and S_3) at incremental distances Δz and time steps Δt as desired. A finite difference from a variable such as C is expressed as:

$$C(z, t) = C(i\Delta z, j\Delta t), \quad i = 1, 2, 3, \ldots, N, \quad and \quad j = 1, 2, 3, \ldots \quad (4\text{-}50)$$

where

$$z = i\Delta z \quad and \quad t = j\Delta t \quad (4\text{-}51)$$

For simplicity the concentration $C(x, t)$ may be abbreviated as:

$$C(z, t) = C_{i,j} \quad (4\text{-}52)$$

where the subscript i denotes incremental distance in the soil and j denotes the time step. We will assume that the concentration distribution at all incremental distances (Δz) is known for time j. We now seek to obtain a numerical approximation of the concentration distribution at time $j + 1$. The convection–dispersion equation 4-24 must be expressed in a finite difference form. For the dispersion and convection terms, the finite difference forms are

$$\Theta D \frac{\partial^2 C}{\partial z^2} = \Theta D \left[\frac{C_{i+1,j+1} - 2C_{i,j+1} + C_{i-1,j+1}}{2(\Delta z)^2} \right] \quad (4\text{-}53)$$

$$+ \Theta D \left[\frac{C_{i+1,j} - 2C_{i,j} + C_{i-1,j}}{2(\Delta z)^2} \right] + O(\Delta z)^2$$

and

$$v \frac{\partial C}{\partial z} = v \left[\frac{C_{i+1,j+1} - C_{i,j+1}}{\Delta z} \right] + O(\Delta z) \quad (4\text{-}54)$$

where $O(\Delta z)^2$ and $O(\Delta z)$ are the error terms associate with the above finite difference approximations, respectively. In eq. 4-53, the second-order derivative (the dispersion term) is expressed in an explicit-implicit form commonly known as the Crank–Nickolson or central approximation method (see Carnahan et al., 1969). This is obtained using Taylor series expansion and is based equally on time j (known) and time $j + 1$ (unknown). Such an approximation has a truncation error, as obtained from the Taylor series expansion, in the order of $(\Delta z)^2$, which is expressed here as $O(\Delta z)^2$. In eq. 4-54, the convection term was expressed in a fully implicit form that resulted in a truncation error of $O(\Delta z)$. In this numerical approximation, for small values of Δz and Δt, these truncation errors were assumed to be sufficiently small and were therefore ignored in our analysis (see Henrici,

1962). Chaudhari (1971) showed that due to numerical approximations, a correction to the dispersion term D is needed such that:

$$D^* = D + D_n \tag{4-55}$$

where D^* is corrected dispersion and D_n is a numerical dispersion term obtained from rearrangement of higher order terms of the Taylor series. For derivatives based on central differences, D_n is given by (Chaudhari, 1971):

$$D^n = \frac{v}{2}\left(\Delta z - \frac{v\Delta t}{R}\right) \tag{4-56}$$

Incorporation of D_n is a simple task and can yield significant improvement to numerical approximations.

The time-dependent term of eq. 4-24 was expressed as:

$$R\frac{\partial C}{\partial t} = R_{i,j}\left[\frac{C_{i,j+1} - C_{i,j}}{\Delta t}\right] + O(\Delta t) \tag{4-57}$$

where the retardation term R was solved explicitly as:

$$R = 1 + \left[\frac{\rho K_f}{\Theta}\right]bC^{b-1} \tag{4-58}$$

This term was also expressed in a finite difference (with iteration) as:

$$R = [R_{i,j}]^r = 1 + \frac{b\rho K_f}{\Theta}[Y^{b-1}]^r \tag{4-59}$$

where Y represents the average concentration at time j (known) and that at time $j + 1$ (unknown) for which solution is being sought such that:

$$[Y]^r = \frac{[C_{i,j+1}]^r + C_{i,j}}{2} \tag{4-60}$$

where r refers to the iteration step. Iteration is implemented here due to the nonlinearity of the equilibrium retention reaction (i.e., $b \neq 1$).

For the kinetic retention equations, the time derivative for S_1, S_2, and S_3 were also expressed in their finite difference forms in a similar manner to the above equations. Therefore, omitting the error terms and incorporating iteration, the term associated with S_1 was expressed as:

$$\rho \frac{\partial S_1}{\partial t} \cong \Theta k_1 \left[\frac{[C_{i,j+1}]^r + C_{i,j}}{2} \right]^n - \rho k_2 [(S_1)_{i,j}]^r \qquad (4\text{-}61)$$

Similarly, the kinetic term associated with S_2, when S_3 is neglected, is approximated as:

$$\rho \frac{\partial S_2}{\partial t} \cong \Theta k_3 \left[\frac{[C_{i,j+1}]^r + C_{i,j}}{2} \right]^m - \rho k_4 [(S_{2_{i,j}}]^r \qquad (4\text{-}62)$$

Moreover, the irreversible term Q was expressed in an implicit-explicit fashion as:

$$Q \cong \Theta k_s \frac{C_{i,j+1} + C_{i,j}}{2} \qquad (4\text{-}63)$$

The number of iterations for the above calculations must be specified since no criteria are commonly given for an optimum number of iterations. A convenient way is based on mass balance calculations (input versus output) as a check on the accuracy of the numerical solution.

For each time step $(j + 1)$, the finite difference of the solute transport equation, after rearrangement and incorporation of the initial and boundary conditions in their finite difference form, can be represented by a set of N equations having N unknown concentrations. The form of the N equations is

$$a_{i,j} C_{i-1,j+1} + d_{i,j} C_{i,j+1} + u_{i,j} C_{i+1,j+1} = e_{i,j} \qquad (4\text{-}64)$$

where N is the number of incremental distances in the soil ($N = L/\Delta x$). The coefficients a, d, u, and e are the associated set of equation parameters. The above N equations were solved simultaneously, for each time step, using the Gaussian elimination method (Carnahan et al., 1969) in order to obtain the concentration C at all nodal points (i) along the soil profile. Solution for a set of linear equations such as the Thomas algorithm for tridiagonal matrix-vector equations can be used (Pinder and Gray, 1977). The newly calculated C values can be subsequently used as input values in the solution for the retention eqs. 4-43 to 4-47. The solution of these equations thus provides the amount of sorbed phases due to the irreversible and reversible reactions at the same time $(j + 1)$ and all incremental distances along the soil profile.

The numerical approximation scheme described above can also be used to solve the solute retention equations associated with the multireaction approach when transport mechanisms are not considered. The solution becomes that representing no-flow batch conditions where the retention is to be described over time. The major exception between the above formulations and that excluding transport is in the way the equilibrium sorbed phase concentration (S_e) is calculated. For any given time step j, the amount in soil solution C and that in the sorbed phase

S_e are in local equilibrium (Rubin, 1983), and their amounts are related by the K_f value according to the nonlinear Freundlich eq. 4-42. Therefore, the total amount in the solution and sorbed phases (S_e) is

$$H = \Theta C + \rho S_e = \Theta C + \rho K_f C^b, \qquad t \geq 0 \qquad (4\text{-}65)$$

As a result, one can calculate, from C and S_e, the amount H at any time step j. Now to estimate these variables at time step ($j + 1$), subsequent to the calculations of all other variables (i.e., S_1, S_2, etc.), one can calculate a new value for H and partition such a value between C and S_e (based on the Freundlich equation) using the following expression:

$$C = \frac{H}{\Theta + \rho K_f C^{b-1}} \qquad (4\text{-}66)$$

which is derived directly from eq. 4-42, which is based on the newly calculated H for the sum of concentration and equilibrium sorbed phases, H. Equation 4-66 is an implicit equation for C and where iteration is necessary. Specifically, a solution for C at each time step j and iteration step r can be obtained as follows:

$$[C_{i,j+1}]^{r+1} = \frac{H}{\Theta + \rho K_f ([C_{i,j+1}]^r)^{b-1}} \qquad (4\text{-}67)$$

All other finite difference expressions can be the same as those used for the multireaction and transport model (eqs. 4-55 to 4-60).

SENSITIVITY ANALYSIS

In order to illustrate the kinetic behavior of heavy metal retention as governed by the multireaction and transport model (MRTM), several simulations are shown. Figures 4-8 through 4-16 are selected simulations illustrating the sensitivity of solution concentration results to a wide range of parameters necessary for the multireaction model, with emphasis on the governing retention mechanism. The parameters selected for the sensitivity analyses are: $\rho = 1.25$ g cm^{-3}, $\Theta = 0.4$ cm^3 cm^{-3}, $L = 10$ cm, $C_i = 0$, $C_o = 10$ mg L^{-1}, and $D = 1.0$ cm^2 h^{-1}. Here we assume a solute pulse was applied to a fully water-saturated soil column initially devoid of solute. In addition, a steady water flow velocity (q) is assumed constant, with a Peclet number P (= $qL/\Theta D$) of 25. The length of the pulse was assumed to be 3 pore volumes, which was then followed by several pore volumes of a solute-free solution.

The influence of the distribution coefficient K_f, which is associated with S_e of the equilibrium type reaction on the transport of dissolved chemicals such as heavy metals, is shown in Figure 4-8. Here the nonlinear parameter b was chosen as 0.5 and all reaction coefficients ($k_1, \ldots, k_6$, and k_s) were set equal to zero. As a result, the shape of the breakthrough curves (BTCs) of Figure 4-8 reflects the influence of nonlinear equilibrium Freundlich-type sorption. For the nonreactive case ($K_f = 0$), which indicates no solute retardation, the sorption (or effluent) side and the desorption side of the BTC appear symmetric. Here the solute concentration (C/C_o) slightly exceeds 0.5 for V/V_0 of 1. As the Freundlich distribution coefficient K_f (eq. 4-42) increased, the solute became more retarded, as is clearly illustrated by the location of the sorption side of the BTCs. For example, for the case where $K_f = 2$, nearly three pore volumes were required prior to the detection of solute in the effluent solution. In the meantime, a reduction of concentration maxima and the presence of tailing of the desorption side is observed for large K_f values. This is due not only to the large K_f values used, but also to the nonlinearity of the equilibrium mechanism ($b \neq 1$) chosen here.

The influence of a wide range of b values on the shape of the BTC is shown in Figures 4-9 and 4-10. For all the BTCs shown in Figures 4-9 and 4-10, a K_f of unity was used, with all other rate coefficients set equal to zero. For values of $b < 1$, the shape of the BTCs indicate a sharp rise in concentration or a steep sorption side with an increase of the tailing of the desorption side for decreasing values of b. In contrast, for $b > 1$, the sorption side indicates a slow increase of concentration, which is associated with a lack of tailing of the desorption side of the BTCs.

The significance of the rate coefficients k_1 and k_2 of the MRTM model on solute retention and transport may be illustrated by the BTCs of Figures 4-11 and 4-12, where a range of rate coefficients differing by three orders of magnitude were chosen. For these simulations, values for k_3 to k_6 and k_s were maintained

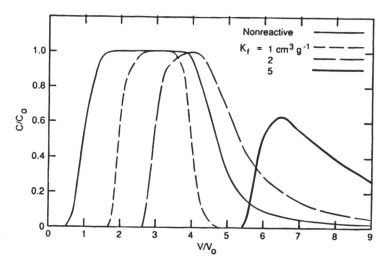

Figure 4-8 Breakthrough curves for several K_f values with $b = 0.5$ and $k_1 = k_2 = \cdots = k_s = 0$.

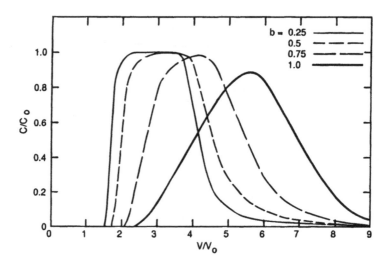

Figure 4-9 Breakthrough curves for several b values where $b \leq 1.0$ and $k_1 = k_2 = \cdots = k_s = 0$.

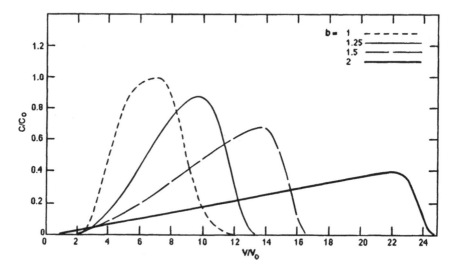

Figure 4-10 Breakthrough curves for several b values where $b \geq 1.0$ and $k_1 = k_2 = \cdots = k_s = 0$.

equal to zero and the equilibrium reaction was assumed linear where $K_f = 1$ cm^3 g^{-1} and b was set equal to unity. For the BTCs shown in Figure 4-11, the forward rate coefficients were constant ($k_1 = 0.10$ h^{-1}), whereas k_2 varied from 1 to 0.001 h^{-1}. A decrease in concentration maxima and a shift of the BTCs resulted as the value for k_2 decreased. Such a shift of the BTCs signifies an increase in solute retention due to the influence of the kinetic mechanism associated with S_1. As the rate of backward reaction (k_2) decreases or k_1/k_2 increases, the amount

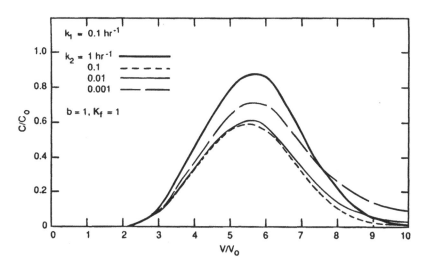

Figure 4-11 Breakthrough curves for several values of the rate coefficients k_1 and k_2 (where $k_3 = k_4 = \cdots = k_s = 0$).

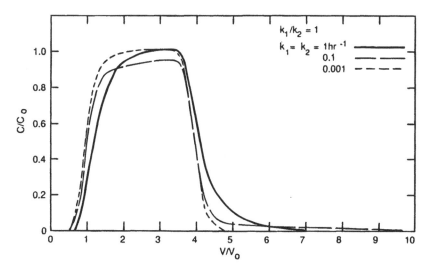

Figure 4-12 Breakthrough curves for several values of k_1 and k_2, (where $k_3 = k_4 = \cdots = k_s = 0$).

of S_1 retained increases and solute mobility in the soil becomes more retarded. The BTCs of Figure 4-12 illustrate the significance of the magnitude of the kinetic rate reactions k_1 and k_2 while the ratio k_1/k_2 remained constant. It is obvious that as the magnitude of the rate coefficients increased, the amount of solute retained increased and an increased solute retardation became evident. Moreover, for extremely small k_1 and k_2 values (e.g., 0.001 h^{-1}), the BTC resembles that for a nonreactive solute due to limited contact time for solute retention by the soil matrix under the prevailing water flow velocity conditions. On the other hand, large rate coefficients are indications of fast or instantaneous-type retention reactions. Specifically, rapid reactions indicate that the retention process is less kinetic and approaches equilibrium conditions in a relatively short contact time.

Figure 4-13 shows BTCs for several values of the nonlinear parameter n, which is associated with the kinetic retention reaction for S_1. There are similar features between BTCs of the previous figures and the simulations illustrated here. An increase in the value of n resulted in a decrease in peak solute concentration and was accompanied by very strong tailing of the BTCs. No apparent shift of the BTCs was observed as n increased from 0.25 to 1.5. Similarities between the influence of n and the nonlinear parameter b associated with the equilibrium reactions are apparent when one compares the BTCs shown in Figure 4-13 with those of Figures 4-9 and 4-10.

In the BTCs shown above, the irreversible retention mechanism for solute removal via the sink term was ignored (i.e., $Q = 0$). The influence of the irreversible kinetic reaction is a straightforward one as shown in Figures 4-14 and 4-15. This is manifested by the lowering of the solute concentration for the overall BTC for increasing values of k_s. Since a first-order irreversible reaction was assumed for the sink term, the amount of irreversibly retained solute (thus lowering of the BTC) is proportional to the solution concentration. The primary difference between the BTCs shown in Figures 4-14 and 4-15 is due to the value of the nonlinear parameter b associated with the equilibrium retention mechanism. In Figure 4-14, b is 0.5, whereas $b = 1$ in Figure 4-15. All other parameters remained invariant and were $k_1 = 0.001$, $k_2 = 0.01$ h^{-1}, $K_f = 1$ cm^3 g^{-1}, and $k_3 = k_4 = k_5 = k_6 = 0$.

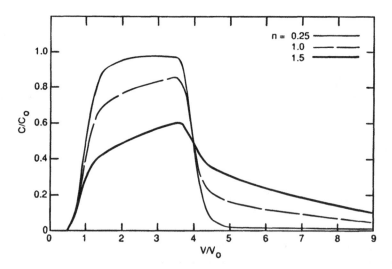

Figure 4-13 Breakthrough curves for several values of the nonlinear parameter n associated with S_1 (where $k_3 = k_4 = \cdots = k_s = 0$).

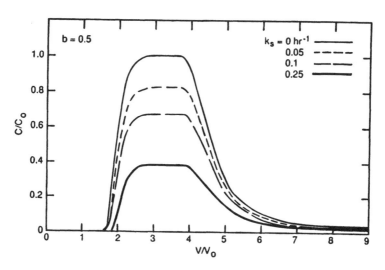

Figure 4-14 Breakthrough curves for several values of the irreversible rate coefficient k_s (where $b = 0.5$ and $k_3 = k_4 = \cdots = k_6 = 0$).

The presence of a consecutive type reaction between S_2 and S_3 in the MRTM model is manifested by the BTCs shown in Figures 4-16 and 4-17. In Figure 4-16, the influence of the magnitude of the rate of reactions k_5 and k_6 is illustrated. As k_5/k_6 increases, little influence on the retardation of the BTCs was observed. However, a decrease in peak concentration and an overall lowering of solute concentration of the desorption side is apparent. The incorporation of the S_3 phase in the model has the distinct advantage that one can assume that such a

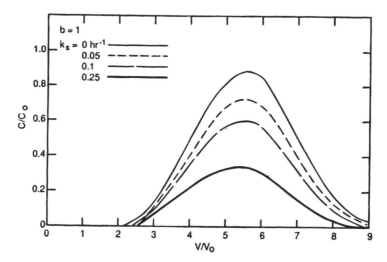

Figure 4-15 Breakthrough curves for several values of the irreversible rate coefficient k_s (where $b = 1$ and $k_3 = k_4 = \cdots = k_6 = 0$).

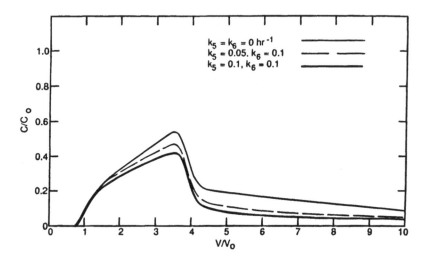

Figure 4-16 Breakthrough curves for several values of the rate coefficients k_5 and k_6 (where $k_3 = k_4 = k_s = 0$).

consecutive type reaction may be regarded as a slow mechanism with slow rate of reaction. Furthermore, if the backward rate is exceedingly small ($k_6 = 0$, Figure 4-17), the consecutive reaction becomes an irreversible mechanism. As a result, in this model, the presence of an irreversible reaction may not be limited to that of the sink term of eq. 4-47 with a direct reaction with the soil solution. Simulations that illustrate the influence of irreversible retention for the consecutive reaction are shown in Figure 4-17.

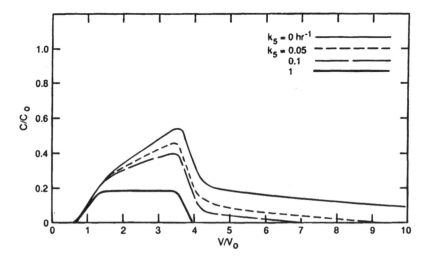

Figure 4-17 Breakthrough curves for several values of the rate coefficient k_5 (where $k_3 = k_4 = k_6 = k_s = 0$).

APPLICATIONS

The capability of the nonlinear multireaction model to describe the transport of heavy metals is illustrated here for Cr(VI) and Cd and for several soils. Breakthrough curves for Cr(VI) transport from miscible displacement experiments are shown in Figures 4-18 to 4-22 for three soils. Details of soil physical parameters and fluxes of the miscible displacement experiments are given in Selim et al. (1989). Results from Webster soil show high peak concentrations close to unity ($C/C_o = 1$), with a sharp rise of the influent side of the BTCs. This was accompanied by little tailing of the desorption side of the BTCs. The times of arrival (or location) of the BTCs also suggest that little retention of Cr(VI) occurred in this soil. For Olivier soil (Figure 4-19), the BTC has a high peak concentration and a moderate tailing of the desorption side. However, there is a shift to the right of the entire BTC, which suggests the occurrence of retardation (sorption/desorption) during transport in Olivier soil. In addition, the BTC results of Figures 4-18 and 4-19 show that for Webster and Olivier soils, approximately 100% of the applied Cr(VI) pulse was recovered in the effluent solutions. For Windsor and Cecil soils (Figures 4-20 to 4-22), total recovery of the applied Cr(VI) pulse was not achieved. The areas under the BTCs were about 60 and 30% of the applied pulses for Windsor and Cecil soils, respectively. The BTCs show extensive tailing and a retardation of Cr(VI) in the effluent as well as a lowering of concentration maximum, which strongly suggests kinetic retention behavior and irreversible or slowly reversible reactions.

The solid and dashed curves shown in Figures 4-18 to 4-22 are calculations using the nonlinear multireaction model of Cr(VI) for the miscible displacement

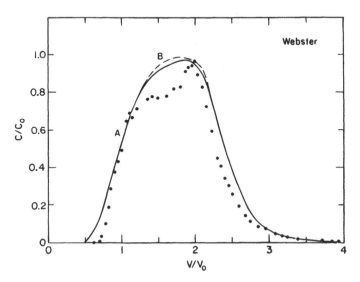

Figure 4-18 Measured (closed circles) and predicted BTCs for Cr(VI) in Webster soil. Curve A is model prediction using batch rate coefficients for $C_o = 10$ mg L^{-1}, and curve B (dashed) is model results for a nonreactive solute.

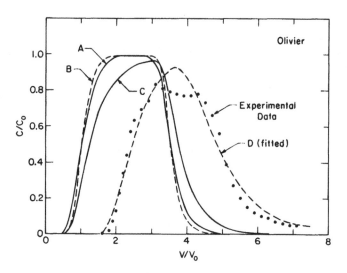

Figure 4-19   Measured (closed circles) and predicted BTCs for Cr(VI) in Olivier soil. Curves
A, B, and C are model predictions using batch rate coefficients for C_o = 100,
10, and 1 mg L^{-1}, respectively. Curve D is a fitted BTC using parameters
obtained from least-squares optimization.

experiments. To obtain such predictions, input model parameters (k_1, k_2, k_3, k_4,
and k_s) derived from the batch reaction studies (Amacher et al., 1988) were used.
To predict Cr(VI) transport in Webster soil (Figure 4-18), the multireaction
approach was used in two different ways. Because of the overall shape and
location of the BTC, a nonreactive model to describe the experimental data was

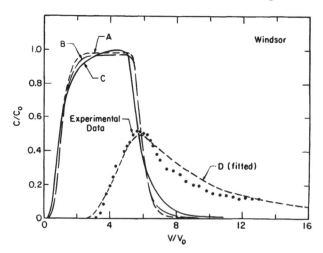

Figure 4-20   Measured (closed circles) and predicted BTCs for Cr(VI) in Windsor soil.
Curves A, B, and C are model predictions using batch rate coefficients for
C_o = 2, 5, and 100 mg L^{-1}, respectively. Curve D is a fitted BTC using
parameters obtained from least-squares optimization.

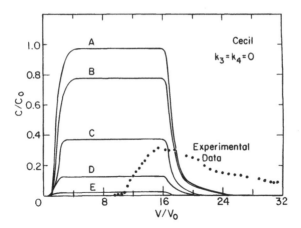

Figure 4-21 Measured (closed circles) and predicted BTCs for Cr(VI) in Cecil soil. Curves A, B, C, D, and E are model predictions using batch rate coefficients for C_o = 100, 25, 5, 2, and 1 mg L^{-1}, respectively. A three-parameter model version was used in all predictions (i.e., $k_3 = k_4 = 0$).

utilized (dashed curve). Model predictions using estimated parameters from the batch results and a three-parameter version of the multireaction model (with $S_2 = S_3 = 0$) were also utilized. Similarities of model predictions (solid curve) to that for the nonreactive BTC (dashed curve) were not surprising. This is primarily due to the large k_1 and k_2 values (0.948 and 1.08 h^{-1}), which suggests rapid retention–release reactions and, thus, closely resemble conditions for local equilibrium. In addition, the extent of Cr(VI) retention was small due to the small values for the order of reaction n (= 0.303) and k_s (= 0.00258 h^{-1}). The model estimated that only 0.1% of applied Cr(VI) was irreversibly retained by the soil column after 4 pore volumes.

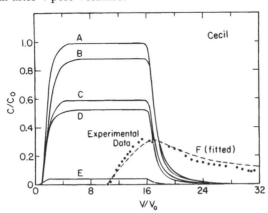

Figure 4-22 Measured (closed circles) and predicted BTCs for Cr(VI) in Cecil soil. Curves A, B, C, D, and E are model predictions using batch rate coefficients for C_o = 100, 25, 5, 2 and 1 mg L^{-1}, respectively. Curve F is a fitted BTC using parameters obtained from least-squares optimization.

Model predictions of Cr(VI) transport for Olivier, Windsor, and Cecil soils are given in Figures 4-19 to 4-22, based on several sets of parameter values for the rate coefficients (k_1, k_2, k_3, k_4, and k_s) in the transport model (Selim et al., 1989). This is because a unique set of values for the rate coefficients were not obtained from the batch data; rather, a strong dependence of rate coefficients on input concentration (C_o) was observed, as discussed in Chapter 3. For these soils, several features of the predicted BTCs are in common and suggest strong dependence on the set of rate coefficients used in model predictions. Increased sorption and decreased peak concentrations were predicted when batch rate coefficients from low initial concentrations (C_o) were used. The use of batch rate coefficients at $C_o = 100$ mg L^{-1}, which is the concentration of Cr(VI) in the input pulse, grossly underestimated Cr(VI) retention by the predicted BTCs for these soils. In fact, underestimation of Cr(VI) retention, and thus overestimation of potential mobility of Cr in these soils, was consistently observed for all sets of rate coefficients used.

The dependence of BTC predictions on the version of the model used in obtaining the necessary rate coefficients is clearly depicted by the predictions shown in Figures 4-21 and 4-22 for Cecil soil. In Figure 4-21, the predictions were obtained using the three parameter model version where $S_2 = S_3 = 0$, whereas the BTCs in Figure 4-22 were obtained using the five-parameter version ($S_3 = 0$). It is evident that the use of these two different model versions resulted in different model predictions as clearly illustrated by the BTCs C and D in Figures 4-21 and 4-22. This is not always the case, however, based on predictions of retention data from the batch experiments (Amacher et al., 1988).

Breakthrough results for Cd transport from a miscible displacement experiment for Windsor soil are shown in Figure 4-23. Details of soil physical parameters and fluxes of other Cd miscible displacement experiments are given in Selim et al. (1992).

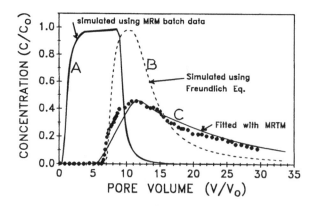

Figure 4-23 Measured (closed circles) and predicted BTCs for Cd in Windsor soil. Curves A and B are model predictions using the multireaction and Freundlich models, respectively. Curve C is a fitted BTC using the multireaction model.

Attempts were made to describe the BTC for Windsor soil based on rate coefficients (k_1, k_2, k_3, k_4, and k_s) obtained independently from batch experiments. Specifically, rate coefficients associated with C_o = 100 mg L^{-1}, which is the concentration of Cd in the input pulse, were used. Predicted results shown by curve A in Figure 4-23 indicate a shift to the left of measured BTC, indicating little retardation with a peak concentration close to unity (C/C_o = 1). Since estimates for the rate coefficients were dependent on input concentrations (C_o) and no unique set of parameter values was obtained, model predictions are also shown for rate coefficients associated with C_o of 10 and 50 mg L^{-1}. It is surprising to find that such model predictions yielded similar results to that of curve A. These results suggest only a weak dependence on the set of rate coefficients used. Therefore, the use of batch rate coefficients grossly underestimated the extent of Cd retention in Windsor soil. Another attempt to describe the BTC of Cd was carried out assuming equilibrium Freundlich sorption as the retention mechanism. The K_f and b parameter values used were based on Cd isotherm results for 24 h of retention. Model prediction is given by curve B in Figure 4-7, where K_f = 28.44 cm^3 g^{-1} and b = 0.701. Model simulations predicted greater retardation than the measured BTC. When K_f and b values from the 240-h retention isotherm were used, a greater shift of predicted results was obtained with a maximum C/C_o of 0.38 at V/V_o = 24 (not shown). It is recognized that the use of the Freundlich (equilibrium) approach is not strictly valid according to the kinetic nature of the Cd results for this soil (see Chapter 3). In fact, overestimation of Cd retention using the Freundlich approach was not surprising since decreased retention during transport is expected when nonequilibrium conditions prevail. On the other hand, reasons for the failure of the batch kinetic rate coefficients to describe measured BTC for the Windsor soil are not completely understood. It is conceivable that the batch data give more emphasis to the forward rate coefficients (k_1 and k_3) than to the backward rates. This is perhaps because sorption in batch experiments is the dominant reaction, especially during early reaction times.

Model predictions shown in Figures 4-18 to 4-23 clearly illustrate the failure of input parameters based on measured batch results to describe Cr(VI) and Cd transport data using the nonlinear multireaction model. The reasons for this failure, which has been observed by other scientists, are not fully understood. A likely explanation is that due to experimental artifacts of the batch procedure compared to prevailing conditions of miscible displacement methods for the transport of solutes in soil columns. Another likely explanation is that the proposed model does not completely account for all reaction mechanisms and/or reaction components present in the soil system. Specifically, the model is perhaps an apparent rather than a complete mechanistic rate law. It is conceivable that the strong dependence of rate coefficients on C_o, as obtained from batch experiments, may also be due to additional mechanisms still ignored in the model. Invariant rate coefficients with solution concentrations are expected if a complete mechanistic rate law is postulated.

A number of scientists have argued that an excellent fit of a data set does not in itself constitute proof of any specific retention/release mechanism (Amacher et al., 1988). One can test the capability for describing (or fitting) transport data for several heavy metals and soils without reliance on input parameter from batch data sets. The hypothesis is that a model gives an inaccurate representation of the reaction mechanism and should thus be discarded if it is completely incapable of describing experimental data sets such as the BTCs shown in Figures 4-18 to 4-23. For all Cr(VI) and Cd cases, the goodness-of-fit, as measured by r^2 exceeded 0.90, and thus the use of the nonlinear multireaction approach, is a predictive tool for describing heavy metals in soils. The model can also be used to provide some interpretations based on description of the data and associated parameters that provided best-fit predictions. As an example, for Olivier soil, the use of a nonlinear (Freundlich) equilibrium rather than kinetic retention reactions provided excellent Cr(VI) BTC prediction with no apparent need for irreversible (sink term) sorption from the soil solution. Another method to find out whether instantaneous equilibrium or kinetic retention is the appropriate modeling approach is by the use of the dimensionless Damkohler number ($D_k = kL/v$), which is dependent on the rate coefficient k as well as the pore water velocity (v/Θ) and column length, L. Jennings (1987) suggested that D_k values above 100 as a critical value for equilibrium reactions of adsorption or ion exchange be used. For Windsor and Cecil soils, the corresponding Damkohler numbers strongly suggest kinetic behavior as the dominant mechanism.

5 A SECOND-ORDER, TWO-SITE TRANSPORT MODEL

In this chapter, an analysis of a kinetic second-order approach for the description of heavy metals retention mechanisms in the soil environment is presented. The basis for this approach is that it accounts for the sites on the soil matrix that are accessible for retention of heavy metals in solution. The second-order approach will be incorporated into the nonequilibrium two-site model for the purpose of simulation of the potential retention during transport of heavy metals in soils. As will be described in the subsequent chapter, this second-order approach will be extended to the diffusion-controlled mobile–immobile (or two-region) transport model.

A main feature of the second-order, two-site model proposed here is the supposition that there exists two types of retention sites on soil matrix surfaces. Moreover, the primary difference between these two types of sites is based on the rate of the proposed kinetic retention reactions. It is also assumed that the retention mechanisms are site-specific; for example, the sorbed phase on type 1 sites may be characteristically different (in their energy of reaction and/or the identity of the solute-site complex) from that on type 2 sites. An additional assumption is that the rate of solute retention reaction is a function of not only the solute concentration present in the solution phase, but also on the amount of available retention sites on matrix surfaces. Another feature of the second-order approach is that an adsorption maxima (or capacity) is assumed. For a specific heavy metal, this maxima represents the total amount of adsorption sites on the soil matrix. This adsorption maximum is also considered an intrinsic property of an individual soil and is thus assumed constant (Selim and Amacher, 1988).

SECOND-ORDER KINETICS

For simplicity, we denote S_{max} to represent the total retention capacity or the maximum amount of adsorption sites on matrix surfaces. It is also assumed that S_{max} is invariant with time. Therefore, based on the two-site approach, the total amount of sites can be partitioned into two types such that:

$$S_{max} = (S_{max})_1 + (S_{max})_2 \tag{5-1}$$

where $(S_{max})_1$ and $(S_{max})_2$ are the total amounts of type 1 sites and type 2 sites, respectively. If F represents the fraction of type 1 sites to the total amount of sites or the adsorption capacity for an individual soil, we have

$$(S_{max})_1 = FS_{max} \quad and \quad (S_{max})_2 = (1 - F)S_{max} \tag{5-2}$$

We now denote ϕ as the amount of unfilled or vacant sites in the soil such that:

$$\phi_1 = (S_{max})_1 - S_1 = FS_{max} - S_1 \tag{5-3}$$

$$\phi_2 = (S_{max})_2 - S_2 = (1 - F)S_{max} - S_2 \tag{5-4}$$

where ϕ_1 and ϕ_2 are amounts of vacant sites and S_1 and S_2 are the amounts of solute retained (or the filled sites) on type 1 and type 2 sites, respectively. As the sites become filled or occupied by the reactive solute, the amount of vacant sites approaches zero, i.e., $(\phi_1 + \phi_2) \to 0$. In the meantime, the amount of solute retained by the soil matrix approaches that of the total capacity (or maximum amount) of sites, $(S_1 + S_2) \to S_{max}$.

We commonly express the amount of heavy metal retained, such as S_1 and S_2 of eqs. 5-3 and 5-4, as the mass of solute per unit mass of soil (mg kg^{-1} soil). Based on the above formulations, the amount of total sites S_{max}, $(S_{max})_1$, and $(S_{max})_2$ and vacant or unfilled sites ϕ_1 and ϕ_2 must also have similar dimensions. Here, the units used, S, and ϕ, will be in terms of milligrams of solute per kilogram soil mass (mg kg^{-1} soil).

Based on this approach, heavy metal retention mechanisms are assumed to follow a second-order kinetic type reaction where the forward process is controlled by the product of the solution concentration C (mg L^{-1}) and the amount of unoccupied or unfilled sites (ϕ) (Selim and Amacher, 1988). Specifically, the reactions for type 1 and type 2 sites may be expressed, respectively, by the reversible processes:

$$C + \phi_1 \underset{k_2}{\overset{k_1}{\rightleftarrows}} S_1 \tag{5-5}$$

and

$$C + \phi_2 \underset{k_4}{\overset{k_3}{\rightleftarrows}} S_2 \tag{5-6}$$

Therefore, the differential form of the kinetic rate equations for heavy metal retention may be expressed, respectively, as:

$$\rho \frac{\partial S_1}{\partial t} = k_1 \Theta \phi_1 C - k_2 \rho S_1 \quad \textit{for type 1 sites} \tag{5-7}$$

and

$$\rho \frac{\partial S_2}{\partial t} = k_3 \Theta \phi_2 C - k_4 \rho S_2 \quad \textit{for type 2 sites} \tag{5-8}$$

where k_1 and k_2 (h^{-1}) are forward and backward rate coefficients for type 1 sites, whereas k_3 and k_4 are rate coefficients for type 2 reaction sites, respectively. In addition, Θ is the soil water content (cm^3 cm^{-3}), ρ is the soil bulk density (g cm^{-3}), and t is time (h). If ϕ_1 and ϕ_2 are omitted from eqs. 5-7 and 5-8, the above equations yield two first-order kinetic retention reactions (Lapidus and Amundson, 1952). However, a major disadvantage of first-order kinetic reactions is that as the concentration in solution increases, a maximum solute sorption is not attained, which implies that there is an infinite solute retention capacity of the soil or an infinite amount of exchange sites on matrix surfaces. In contrast, the approach proposed here achieves maximum sorption when all unfilled sites become occupied (i.e., ϕ_1 and $\phi_2 \rightarrow 0$).

In a fashion similar to the nonequilibrium two-site concept proposed by Selim et al. (1976), it is possible to regard type 1 sites as those where equilibrium is rapidly reached (i.e., in a few minutes or hours). In contrast, type 2 sites are highly kinetic and may require several days or months for apparent local equilibrium to be achieved. Therefore, for type 1 sites, the rate coefficients k_1 and k_2 are expected to be several orders of magnitude larger than k_3 and k_4 of the type 2 sites. As $t \rightarrow \infty$, i.e., when both sites achieve local equilibrium, eqs. 5-7 and 5-8 yield the following expressions. For type 1 sites:

$$k_1 \Theta \phi_1 C - k_2 \rho S_1 = 0, \quad or \quad \frac{S_1}{\phi_1 C} = \frac{\Theta}{\rho} \frac{k_1}{k_2} = \omega_1 \tag{5-9}$$

and for type 2 sites:

$$k_3 \Theta \phi_2 C - k_4 \rho S_2 = 0, \quad or \quad \frac{S_2}{\phi_2 C} = \frac{\Theta}{\rho} \frac{k_3}{k_4} = \omega_2 \tag{5-10}$$

Here, ω_1 and ω_2 represent equilibrium constants for the retention reactions associated with type 1 and type 2 sites, respectively. The formulation of eqs. 5-9 and 5-10 is analogous to expressions for homovalent ion-exchange equilibrium reactions. In this sense, the equilibrium constants ω_1 and ω_2 resemble the selectivity coefficients for exchange reactions, and S_{max} resembles the exchange capacity (CEC) of soil matrix surfaces (Sposito, 1981). However, a major difference between ion exchange and the proposed second-order approach is that no consideration of other competing ions in solution or on matrix surfaces is incorporated into the second-order rate equations. In a strict thermodynamic sense, eqs. 5-9 and 5-10 should be expressed in terms of activities rather than concentrations. However, we use the implicit assumption that solution-phase ion activity coefficients are constant in a constant ionic strength medium. Moreover, the solid-phase ion activity coefficients are assumed to be incorporated into the selectivity coefficients (ω_1 and ω_2) as in ion-exchange formulations (Sposito, 1981).

Incorporation of eq. 5-1 to 5-4 into eq. 5-9 and 5-10 and further rearrangement yields the following expressions for the amounts retained by type 1 and 2 sites at $t \to \infty$:

$$\frac{S_1}{(S_{max})_1} = \left[\frac{\omega_1 C}{1 + \omega_1 C}\right], \quad and \quad \frac{S_2}{(S_{max})_2} = \left[\frac{\omega_1 C}{1 + \omega_2 C}\right] \qquad (5\text{-}11)$$

Therefore, the total amount sorbed in the soil, S, ($= S_1 + S_2$), is

$$\frac{S}{S_{max}} = \left[\frac{\omega_1 C}{1 + \omega_1 C}\right]F + \left[\frac{\omega_2 C}{1 + \omega_2 C}\right](1 - F) \qquad (5\text{-}12)$$

Equation 5-12 is analogous to the two-site Langmuir formulation where the amount sorbed in each region is clearly expressed. Such Langmuir formulations are commonly used to obtain independent parameter estimates for S_{max} and the affinity constants ω_1 and ω_2 (Sposito, 1984).

Let us now consider the case where only one type of active sites is dominant in the soil system. In a similar fashion to the formulations of eqs. 5-9 and 5-10, the kinetics of the reaction can be described by the following equation:

$$\rho \frac{\partial S}{\partial t} = k_f \Theta \phi C - k_b \rho S \qquad (5\text{-}13)$$

Here, k_f and k_b (h^{-1}) are the forward and backward retention rate coefficients, respectively, and S is the total amount of solute retained by the soil matrix surfaces. This formulation is often referred to as the kinetic Langmuir equation. In fact, reaction (eq. 5-13) at equilibrium obeys the widely recognized Langmuir isotherm equation:

$$\frac{S}{S_{max}} = \frac{\omega C}{1 + \omega C} \qquad (5\text{-}14)$$

where ω ($= \Theta k_f / \rho k_b$) is equivalent to that of eq. 5-9 and 5-10. For a discussion on the formulation of the kinetic Langmuir equation, see Rubin (1983) and Jennings and Kirkner (1984).

It should be recognized that the unfilled or vacant sites (ϕ) in eq. (5-7), (5-8), and (5-13) are not strictly vacant. They are occupied by hydrogen, hydroxyl, or other nonspecifically (e.g., Na, Ca, Cl, and NO_3) or specifically (e.g., PO_4, AsO_4, and transition metals) adsorbed species. Vacant or unfilled refers to vacant or unfilled by the specific solute species of interest. The process of occupying a vacant site by a given solute species actually is one of replacement or exchange of one species for another. However, the simplifying assumption on which this model is based is that the filling of sites by a particular solute species need not consider the corresponding replacement of species already occupying the sites. The Langmuir-type approach considered here (eqs. 5-11 to 5-14) is a specialized case of an ion exchange formulation (ElPrince and Sposito, 1981). Alternatively, the competitive Langmuir approach may be used if the identities of the replaced solute species are known (Jennings and Kirkner, 1984; Jennings, 1987).

TRANSPORT MODEL

Incorporation of the second-order two-site reactions into the classical (convection–dispersion) transport equation yields (Brenner, 1962; Nielsen et al., 1986):

$$\Theta \frac{\partial C}{\partial t} + \rho \left(\frac{\partial S_1}{\partial t} + \frac{\partial S_2}{\partial t} \right) = \Theta D \frac{\partial^2 C}{\partial x^2} - q \frac{\partial C}{\partial x} - Q \qquad (5\text{-}15)$$

where D is the hydrodynamic dispersion coefficient ($cm^2\ h^{-1}$), q is Darcy's water flux ($cm\ h^{-1}$), and x is depth (cm). Here, the term Q is a sink representing the rate of irreversible heavy metal reactions by direct removal from the soil solution ($mg\ h^{-1}\ cm^{-3}$). In this model, the sink term was expressed in terms of a first-order irreversible reaction for reductive sorption, precipitation, or internal diffusion as outlined by Amacher et al. (1986, 1988):

$$Q = \Theta k_s C \qquad (5\text{-}16)$$

where k_s is the rate constant for irreversible reaction (h^{-1}). Equation 5-16 is similar to that for diffusion-controlled precipitation reaction if one assumes that the equilibrium concentration for precipitation is negligible and that k_s is related to the diffusion coefficient (Stumm and Morgan, 1981).

For convenience, we define the dimensionless variables:

$$X = \frac{x}{L} \qquad (5\text{-}17)$$

$$T = \frac{qt}{L\Theta} \qquad (5\text{-}18)$$

$$c = \frac{C}{C_o} \qquad (5\text{-}19)$$

$$s = \frac{S}{S_{max}} \tag{5-20}$$

$$\Phi = \frac{\phi}{S_{max}} \tag{5-21}$$

$$P = \frac{qL}{D\Theta} \tag{5-22}$$

where T is dimensionless time, equivalent to the number of pore volumes leached through a soil column of length, L, and P is the Peclet number (Brenner, 1962). Given the above variables, eqs. 5-15, 5-7, and 5-8 are rewritten in dimensionless form, respectively, as (Selim and Amacher, 1988):

$$\frac{\partial c}{\partial T} + \Omega \left[\frac{\partial s_1}{\partial T} + \frac{\partial s_2}{\partial T} \right] = \frac{1}{P} \frac{\partial^2 c}{\partial X^2} - \frac{\partial c}{\partial X} - \kappa_s c \tag{5-23}$$

$$\frac{\partial s_1}{\partial T} = \kappa_1 \Phi_1 c - \kappa_2 s_1 \tag{5-24}$$

$$\frac{\partial s_2}{\partial T} = \kappa_3 \Phi_2 c - \kappa_4 s_2 \tag{5-25}$$

$$\Omega = \frac{S_{max}\rho}{C_o \Theta} \tag{5-26}$$

$$\kappa_s = \frac{k_s \Theta L}{q} \tag{5-27}$$

$$\kappa_1 = \frac{k_1 \Theta^2 C_o L}{\rho q} \quad and \quad \kappa_3 = \frac{k_3 \Theta^2 C_o L}{\rho q} \tag{5-28}$$

$$\kappa_2 = \frac{k_2 \Theta L}{q} \quad and \quad \kappa_4 = \frac{k_4 \Theta L}{q} \tag{5-29}$$

Here, κ_s, κ_1, κ_2, κ_3, and κ_4 are dimensionless kinetic rate coefficients that incorporate q and L. As will be shown in a later section, these dimensionless variables (including Ω, c, s, and Φ) represent a convenient way to study the sensitivity of the model to reduced variables.

For the purpose of simulation and model evaluation, the appropriate initial and boundary conditions were as follows. We chose uniform initial solute concentration C_i in a finite soil column of length, L, such that:

$$c = c_i \quad (T = 0 \quad and \quad 0 < X < 1) \tag{5-30}$$

We also assume that an input solute–solution pulse having a concentration, C_o, was applied at the soil surface for a (dimensionless) time, T_p, and was then followed by a solute-free solution. As a result, at the soil surface, the following third-type boundary conditions were used (Selim and Mansell, 1976; Parker and van Genuchten, 1984):

$$1 = c - \frac{1}{P} \frac{\partial c}{\partial X} \qquad (X = 0, T < T_p) \qquad (5\text{-}31)$$

$$0 = c - \frac{1}{p} \frac{\partial c}{\partial X} \qquad (X = 0, T > T_p) \qquad (5\text{-}32)$$

and at $x = L$, we have

$$\frac{\partial c}{\partial X} = 0 \qquad (X = 1, T > 0) \qquad (5\text{-}33)$$

The differential equations of the second-order, two-site model described above are of the nonlinear type and analytical solutions are not available. Therefore, eqs. 5-21 to 5-23 must be solved numerically. A finite difference approximation (explicit–implicit) subject to the above initial and boundary conditions was carried out by Selim and Amacher (1988) and presented in Selim et al. (1990).

SENSITIVITY ANALYSIS

In order to illustrate the kinetic behavior of solute retention as governed by the proposed second-order reaction, several simulations were performed. We assumed a no-flow condition in order to describe time-dependent batch (sorption–desorption) experiments. The problem becomes an initial-value problem where closed-form solutions are available.

Kinetics

The retention results shown in Figure 5-1 illustrate the influence of the rate coefficients (k_1 and k_2) on the shape of sorption isotherms (S vs. C). The parameters chosen were those of a soil initially devoid of solute ($C_i = S_i = 0$ at $t = 0$) and a soil-to-solution ratio (ρ/Θ) of 1:10, which is commonly used in batch experiments. Since the amount sorbed was assumed to be initially zero, larger values for k_2 than k_1 were selected in our simulations to induce reverse (desorption) reactions.

As shown in Figure 5-1, after 2 days of reaction, isotherm A (where k_1 and k_2 were 0.01 and 0.1 day^{-1}, respectively) appears closer to the equilibrium isotherm than other cases shown. The equilibrium case was calculated using eq. 5-9 and represents an isotherm at t $\rightarrow \infty$, or for a soil having values of k_1 and k_2 that are extremely large. Isotherms B and C represent cases where both k_1 and k_2 values were reduced in comparison to those for isotherm A, by one and

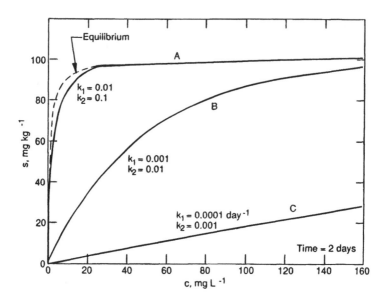

Figure 5-1 Effect of rate coefficients on sorption isotherms using the second-order kinetic model.

two orders of magnitude, respectively. For both cases, the isotherms deviate excessively from the equilibrium case. It is apparent from curve C that 2 days of reaction is insufficient to attain equilibrium and a sorption maxima is not apparent from the shape of the isotherm. Moreover, it is perhaps possible to consider a linear-type isotherm for the concentration range shown. However, as much as 100 days or more of reaction time is necessary to achieve apparent equilibrium conditions. This is illustrated in Figure 5-2, where the influence of time of reaction using the second-order model is shown.

The influence of the sorption maxima (S_{max}) on the retention isotherms is shown in Figure 5-3. The parameters selected were similar to those of Figure 5-2 except that a contact time of 10 days was chosen. As expected, the isotherms reached their respective maxima at lower C values with decreasing S_{max}. The results of Figure 5-3 also indicate a steep gradient of the retention isotherms in the low concentration range. Such a retention behavior has been observed by several scientists for a number of reactive solutes. The simulations also illustrate clearly the influence of the sorption maxima on the overall shape of the isotherms. The influence of other parameters such as F, k_3, and k_4 on retention kinetics can be easily deduced and is thus not shown.

Transport

Figures 5-4, 5-5, and 5-6 are selected simulations that illustrate the transport of a reactive solute with the second-order two-site model as the governing retention mechanism. The parameters selected for the sensitivity analysis were

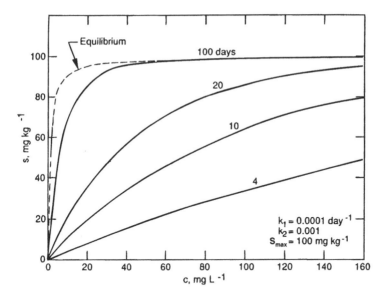

Figure 5-2 Effect of time of retention on sorption isotherms using the second-order kinetic model.

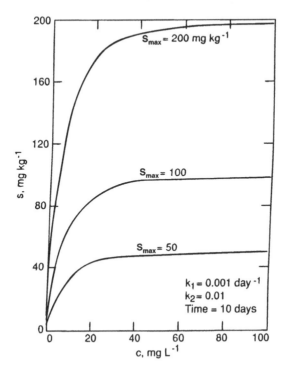

Figure 5-3 Effect of total amount of sites (S_{max}) on the shape of sorption isotherms using the second-order kinetic model.

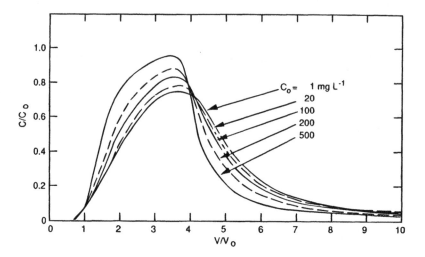

Figure 5-4   Effluent concentration distributions for different initial concentrations (C_o) using the second-order, two-site model.

$\rho = 1.25$ g cm^{-3}, $\Theta = 0.4$ cm^3 cm^{-3}, $L = 10$ cm, $C_i = 0$, $C_o = 10$ mg L^{-1}, $F = 0.50$, and $S_{max} = 200$ mg kg^{-1}. Here we assumed a solute pulse was applied to a fully water-saturated soil column initially devoid of a particular heavy metal of interest. In addition, a steady water flow velocity (q) was maintained constant with a Peclet number, P ($= q/\Theta D$), of 25. The length of the pulse was assumed to be 3 pore volumes, which was then followed by several pore volumes of a heavy metal-free solution. The rate coefficients selected were 0.01, 0.1, 0.001,

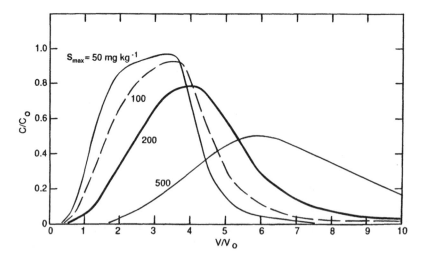

Figure 5-5   Effluent concentration distributions for different S_{max} values using the second-order, two-site model.

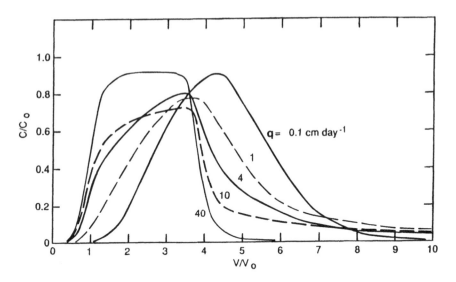

Figure 5-6 Effluent concentration distributions for different flux (q) values using the second-order, two-site model.

and 0.01 day^{-1} for k_1, k_2, k_3, k_4, respectively. As a result, the equilibrium constants ω_1 and ω_2 for sites 1 and 2, respectively, were identical.

Figure 5-4 shows breakthrough curves (BTCs), which represent the relative effluent concentration (C/C_o) vs. *effluent pore volume* (V/V_o), for several input C_o values. The shape of the BTCs is influenced by the input solute concentration and is due to the nonlinearity of the proposed second-order retention mechanism. The simulated results also indicate that for high C_o values, the BTCs appear less retarded and have sharp gradients on the desorption (or right) side. In contrast, for low C_o values, the general shape of the BTCs appears to be kinetic in nature. Specifically, as C_o decreases, a decrease in maximum or peak concentrations and extensive tailing of the desorption side of the BTCs can be observed.

The influence of the total amount of (active) sites (S_{max}) on the BTCs is clearly illustrated by the cases given in Figure 5-5. Here, the value of C_o was chosen constant ($C_o = 10$ mg L^{-1}). The BTCs show that an order of magnitude increase in S_{max} (from 50 to 500 mg kg^{-1}) resulted in an approximately 3 pore volume shift in peak position. In addition, for high S_{max} values, extensive tailing as well as an overall decrease of effluent concentrations (C/C_o) was observed.

The influence of the flow velocity (q) on the shape of the BTC is somewhat similar to that of the rate coefficients for retention provided that the Peclet number (P) remains constant. This is illustrated by the simulations shown in Figure 5-6 for a wide range of flow velocities. For $q = 40$ cm day^{-1}, the retention reactions associated with type 1 sites were not only dominant, but also closer to local equilibrium than those for type 2 sites (results not shown). This is a direct consequence of the limited solute residence time encountered when the fluid flow velocity is exceedingly high. Type 2 sites, which may be considered highly kinetic, were far removed from equilibrium such that only a limited amount of solute was retained from the soil solution. Under such conditions, the amount of available sites (ϕ_2) remains high and the retention capacity of the soil matrix is therefore not achieved. In fact, for $q = 40$ cm day^{-1}, the BTC closely describes a one-site retention mechanism, as indicated by the low retardation and lack of tailing of the desorption side. As the flow velocity decreases, the solute residence time increases and more time is available for the highly kinetic type two sites to retain solute species from the soil solution. In addition, for extremely small velocities ($q \to 0$), the BTC should indicate maximum solute retention during transport. This probably resembles the BTCs with $q = 0.1$ cm day^{-1}, which indicates the highest solute retardation shown. For intermediate velocities (q from 1 to 10), however, the respective BTCs indicate relatively moderate degrees of retardation as well as tailing, which is indicative of kinetic retention mechanisms.

In the BTCs shown in Figures 5-4, 5-5, and 5-6, the irreversible retention mechanism for heavy metal removal (via the sink term) was ignored. The influence of the irreversible kinetic reaction (e.g., precipitation, etc.) is a straight-forward one, as shown in Figure 5-7. This is manifested by the lowering of solute concentration for the overall BTC for increasing values of k_s. Since a first-order reaction was assumed, the lowering of the BTC is proportional to the solution concentration.

In previous BTCs, the sensitivity of model predictions (output) of the second-order approach to selected parameters was discussed. It is convenient, however, to carry out model sensitivity using dimensionless parameters such as those defined by eqs. 5-17 to 5-22. The use of dimensionless parameters offers a distinct advantage over the use of conventional parameters since they provide a wide range of application as well as further insight on predictive behavior of the model.

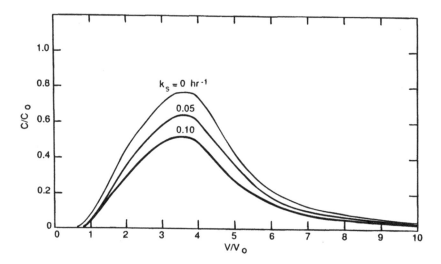

Figure 5-7   Effluent concentration distributions for values of the irreversible rate coefficient (k_s) using the second-order, two-site model.

Figures 5-8 to 5-10 are simulations that illustrate the transport of a reactive solute with the second-order two-site model for selected dimensionless parameters. Unless otherwise indicated, the values for the dimensionless parameters Ω, κ_1, κ_2, κ_3, κ_4, F, κ_s, P and T_p were as follows: 5, 1, 1, 0.1, 0.1, 0, 0.5, 25, and 1, respectively.

Figure 5-8 shows breakthrough curves (BTCs) of a reactive solute for several values of Ω. The figure indicates that the shape of the BTCs is drastically influenced by the value of Ω. This is largely due to the nonlinearity of the proposed second-order retention mechanism. As given by eq. 5-26, Ω represents the ratio of total sites (S_{max}) to input (pulse) solute concentration (C_o). Therefore, for small Ω values (e.g., $\Omega = 0.1$), the simulated BTC is very similar to that for a nonretarded solute due to the limited amount of sites (S_{max}) in comparison to C_o. In contrast, large values of Ω resulted in BTCs that indicate increased retention as manifested by the right shift of peak concentration of the BTCs. In addition, for high Ω values, extensive tailing as well as an overall decrease of effluent concentration was observed.

The effect of the dimensionless reaction rate coefficients (κ_1, κ_2, κ_3, and κ_4) of the two-site model on solute retention and transport is illustrated by the BTCs of Figure 5-9 where a range of rate coefficients differing by three orders of magnitude was chosen. For BTCs shown, the rate coefficients for type 2 sites were chosen to be one order of magnitude smaller than those associated with type 1 sites. These BTCs indicate that depending on the values of κ_1, κ_2, κ_3, and κ_4, two extreme cases can be illustrated. For large values of the κ, rapid sorption–desorption reactions occurred for both type 1 and type 2 sites. Rapid reactions indicate that the retention process is less kinetic and BTCs can approximate local equilibrium conditions in a relatively short contact time. Examples

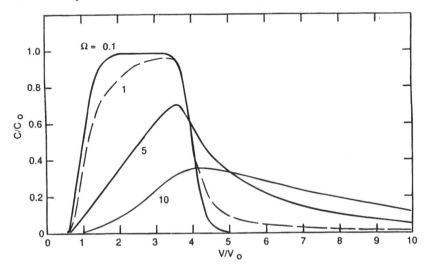

Figure 5-8 Effluent concentration distributions for different values of the parameter Ω of the second-order, two-site model.

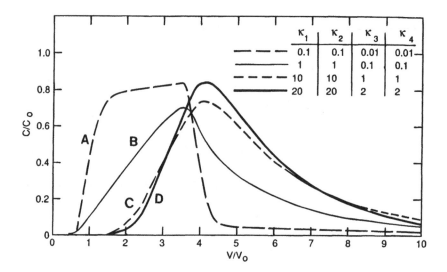

Figure 5-9 Effluent concentration distributions for different values of rate coefficients (κ_1, κ_2, κ_3, and κ_4) using the second-order, two-site model.

are those of curves C and D. In contrast, for extremely small values of κ_1, κ_2, κ_3, and κ_4 (or small residence time), little retention takes place and the shape of the BTC resembles that for a nonreactive solute (see curve A). The behavior of all illustrated BTCs is consistent with those for first-order kinetic and for two-site nonlinear equilibrium–kinetic reactions.

Figure 5-10 shows BTCs for several values of the fraction of sites parameter F. There are similar features between BTCs of Figures 5-8 and 5-9 and those

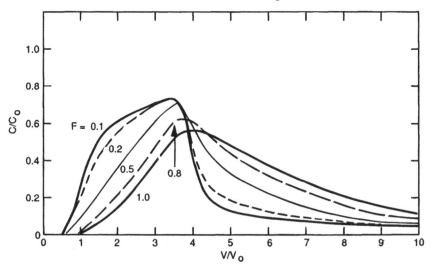

Figure 5-10 Effluent concentration distributions for different values of the fraction of sites F using the second-order, two-site model.

illustrated here. For $F = 1$, all the sites are type 1 sites, which we designated earlier as those sites of strong kinetic influence due to their small values of κ_1 and κ_2. As the contribution of type 2 sites increases (or F decreases), the shape of the BTCs becomes increasingly less kinetic with significant decreases in amount of solute retention.

APPLICATIONS

Application of the second-order two-site model was carried out by Selim and Amacher (1988) and Amacher and Selim (1994) using batch and miscible displacement data sets for Cr(VI) from three different soils. Furthermore, they attempted, whenever possible, to utilize parameters that were either independently measured or estimated by indirect means. The parameters estimated included the adsorption maximum S_{max}, the fraction of sites F, and the kinetic rate coefficients associated with the reversible and irreversible mechanisms.

Estimation of S_{max} and F

Retention data sets for Cr(VI) by three soils (Cecil, Olivier, and Windsor soils) after 336 h of reaction (Figure 5-11) were used to arrive at independent

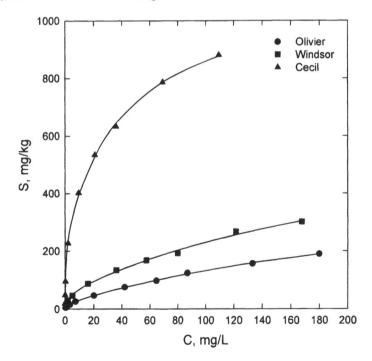

Figure 5-11 Two-site Langmuir sorption curves for Cr(VI) retention by Olivier, Windsor, and Cecil soils after 336 h of reaction.

estimates of S_{max} and F. Specifically, the two-site Langmuir eq. (5-12) was used to describe Cr(VI) retention results using a nonlinear, least-squares, parameter–optimization (curve-fitting) scheme (van Genuchten, 1981). It was assumed that the reactions between Cr(VI) in solution and the two types of sites had attained equilibrium in 336 h even though small amounts of Cr(VI) were still being retained by the soil. The continuing reaction between Cr(VI) in solution and the soil was ascribed to an irreversible reaction included in the model (eq. 5-15). The process responsible for this irreversible reaction is discussed in Chapter 8. It is important to realize that only the reactions of Cr(VI) with the two types of reaction sites were assumed to attain equilibrium in 336 h. Overall retention had not reached equilibrium because of the irreversible reaction. However, if the magnitude of the irreversible term is small, as is the case here, then reliable estimates of S_{max} and F can be made, although the actual S_{max} is somewhat smaller. The statistical results from the parameter optimization scheme (Table 5-1) indicated a close approximation of the two-site Langmuir equation to the experimental sorption isotherms shown in Figure 5-11. However, a close approximation of the data does not constitute proof that two types of reaction sites actually exist (Sposito, 1982). Parameter optimization merely provides a convenient method for estimating retention parameters given in a prescribed model. No reliable and independent experimental method has been developed by which the two types of reaction sites can be distinguished unambiguously and their concentrations accurately measured in soils.

Reaction Kinetics

Data sets of time-dependent retention of Cr(VI) by Olivier, Windsor, and Cecil soils and for several input (initial) concentrations (C_o) were used to provide estimates of the kinetic rate coefficients of the second-order approach (Selim and Amacher, 1988). Both S_{max} and F values previously obtained were used as model inputs and nonlinear parameter–optimization was used to estimate k_1, k_2, k_3, k_4, and k_s. Selim and Amacher (1988) used two versions of the second-order two-site model, a three-parameter or a one-site version (k_1, k_2, and k_s) in which S_{max} was not differentiated into type 1 and type 2 sites ($F = 1$), and a five-parameter or a two-site version (k_1, k_2, k_3, k_4, and k_s) in which two types of reaction sites were considered. For most C_o, either the three- or five-parameter versions described the data adequately, with high r^2 values and low parameter

TABLE 5-1 Two-Site Langmuir Parameters F and S_{max} for Cr(VI) Retention by Olivier, Windsor, and Cecil Soils

Soil	r^2	F	S_{max} (mg kg^{-1})
Olivier	0.9997	0.048 ± 0.007[a]	475 ± 10
Windsor	0.998	0.08 ± 0.02	734 ± 41
Cecil	0.9996	0.22 ± 0.02	1127 ± 24

[a] Parameter value ±95% confidence interval.

From Amacher, M.C. and H.M. Selim. 1997. *Ecol. Mod.* 74:205–230.

standard errors (Selim and Amacher, 1988). The exception was the description of retention data for Olivier soil at high C_o where the retention of Cr(VI) was not highly kinetic and more scatter in experimental data was observed. Model calculations and data for the Olivier, Windsor, and Cecil soils are shown in Figures 5-12, 5-13, and 5-14, respectively.

If the fraction of type 1 sites is small, as was the case with the Olivier and Windsor soils, then their contribution to the kinetic solute retention curve will be small and indistinguishable at high solute concentrations. For the Cecil soil where the fraction of type 1 sites was significant ($F = 0.224$), the five-parameter model version was superior to the three-parameter version at all C_o except for $C_o = 1$ mg L^{-1} (Selim and Amacher, 1988). Therefore, the applicability of the five-parameter version to a wide range of solute concentrations was directly related to the magnitude of the fraction of type 1 sites. As F increased (Olivier < Windsor < Cecil), the concentration range over which the five-parameter version provided a better description of the data than the three-parameter version increased. The shapes of the experimental curves and model calculations (Figures 5-12, 5-13, and 5-14) are influenced by C_o. At higher C_o, retention of Cr(VI) from solution was far less kinetic than at lower C_o. This behavior is as expected if the concentration of one or more reaction sites limits reaction rates. At $C_o =$

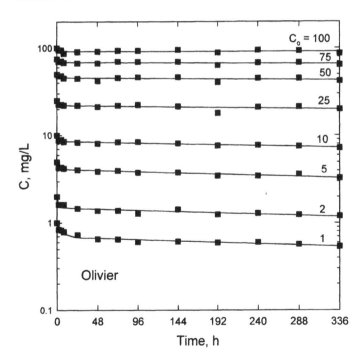

Figure 5-12 Time-dependent retention of Cr(VI) by Olivier soil. Closed squares are the data points and solid lines are second-order, two-site model predictions for different initial concentration curves (C_o = 1, 2, 5, 10, 25, 50, 75, and 100 mg L^{-1}).

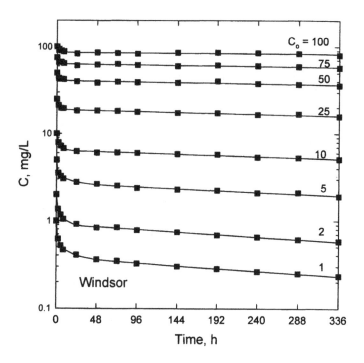

Figure 5-13 Time-dependent retention of Cr(VI) by Windsor soil. Closed squares are the data points and solid lines are second-order, two-site model predictions for different initial concentration curves (C_o = 1, 2, 5, 10, 25, 50, 75, and 100 mg L^{-1}).

100 mg L^{-1}, there was 4 mg Cr(VI) available for reaction in the 40-mL solution volume used in the experiment. The maximum possible amounts of Cr(VI) that could be sorbed by the 4 g of each soil used in the experiment were 1.9, 2.9, and 4.5 mg (solute weight basis) for Olivier, Windsor, and Cecil soils, respectively, based on the S_{max} values of Table 5-1. Thus, maximum possible Cr(VI) retention in the Cecil soil was about equal to the amount of Cr(VI) available for retention, but was much less in the Olivier and Windsor soils than the amount of Cr available. Since the amount of type 1 sites was much less than the total, their contribution to the overall reaction is actually quite negligible at high solute concentrations. The influence of S_{max}/C_o or Ω ($\Omega = \rho S_{max}/C_o\Theta$) on solute retention during transport was illustrated in Figure 5-8.

Although the second-order two-site model can accurately describe the data sets as shown in Figures 5-12, 5-13, and 5-14, this does not mean that it is the correct or only explanation of the data. The model does not depend on any particular process of solute retention. Any or all may be operative, including physical adsorption, formation of outer- or inner-sphere surface complexes, ion exchange, surface precipitation, etc. Furthermore, subsequent solute transformations on the soil surface or internal diffusion into soil particles may occur. Alternative processes may produce the same experimental observations and many

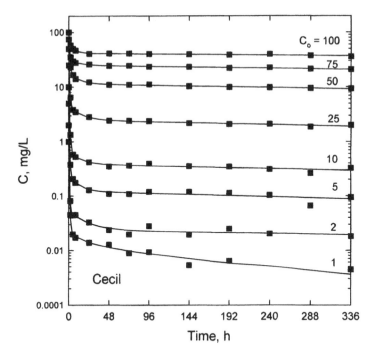

Figure 5-14 Time-dependent retention of Cr(VI) by Cecil soil. Closed squares are data points and solid lines are second-order, two-site model predictions for different initial concentration curves (C_o = 1, 2, 5, 10, 25, 50, 75, and 100 mg L^{-1}).

models may give similar results. This point was thoroughly discussed by Skopp (1986) in his review of time-dependent processes in soils. Amacher et al. (1988) found that a nonlinear multireaction model (see Chapter 4), that does not include concentrations of reaction sites in its formulation, can describe the experimental data as well as the second-order two-site model. Like the second-order model, the nonlinear model is an incomplete description of the actual reactions because the rate coefficients are dependent on C_o. Thus, both types of models yield pseudo rate coefficients.

Transport

Chromium BTCs from miscible displacement experiments for all three soils are shown in Figures 5-15, 5-16, and 5-17. For Cecil and Windsor soils, the measured BTCs appear to be highly kinetic with extensive tailing. For Olivier soil, little tailing was observed and approximately 100% of the applied Cr(VI) pulse was recovered. These results are consistent with the batch data where the irreversible reaction parameter (k_s) was found to be quite small.

In order to examine the capability of the second-order two-site (SOTS) model, the necessary model parameters must be provided. Values for the dispersion coefficients (D) were obtained from BTCs of tracer data for ^3H_2O and ^{36}Cl

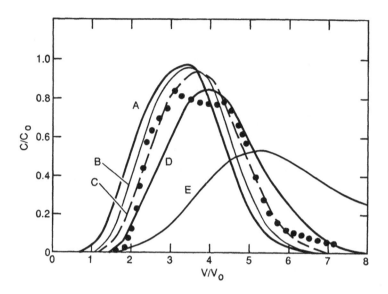

Figure 5-15 Effluent concentration distributions for Cr(VI) in Olivier soil. Curves A, B, C, D, and E are predictions using the second-order, two-site model with batch rate coefficients for C_o of 100, 25, 10, 5, and 1 mg/L, respectively.

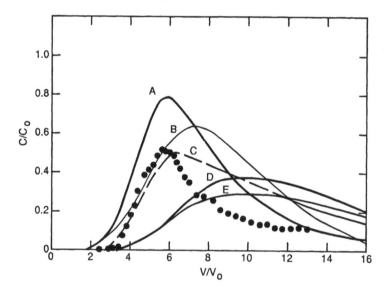

Figure 5-16 Effluent concentration distributions for Cr(VI) in Windsor soil. Curves A, B, C, D, and E are predictions using the second-order, two-site model with batch rate coefficients for C_o of 25, 10, 5, 2, and 1 mg/L, respectively.

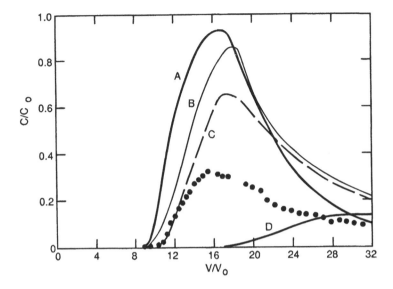

Figure 5-17 Effluent concentration distributions for Cr(VI) in Cecil soil. Curves A, B, C, and D are predictions using the second-order, two-site model with batch rate coefficients for C_o of 100, 50, 25, and 10 mg/L, respectively.

(Chapter 4). Other model parameters such as ρ, Θ, and q were measured for each soil column. In addition, values for S_{max}, and F used in describing Cr (VI) BTCs, using the SOTS model, were those derived from adsorption isotherms shown in Figure 5-11. Direct measurement of these parameters by other than parameter optimization techniques is not available. Moreover, we utilized the reaction rate coefficients k_1, k_2, k_3, k_4, and k_s as obtained from the batch kinetic data in the predictions of Cr(VI) BTCs. In the following discussion, predicted curves imply the use of independently measured parameters in model calculations, as was carried out here using the batch-derived parameters.

The predicted BTCs shown in Figures 5-15, 5-16, and 5-17 were obtained using different sets of rate coefficients (k_1, k_2, k_3, k_4, and k_s) in the SOTS model. This is because no single or unique set of values for these rate coefficients were obtained from the batch data; rather, a strong dependence of rate coefficients on input concentration (C_o) was observed. For all soils, several features of the predicted BTCs are similar and indicate strong dependence on the rate coefficients used in model calculations. Increased sorption during transport, lowering of peak concentrations, and increased tailing were predicted when batch rate coefficients from low initial concentrations (C_o) were used.

From Figures 5-15 to 5-17, the use of batch rate coefficients at $C_o =$ 100 mg L^{-1}, which is the concentration of Cr(VI) in the input pulse, grossly underestimated Cr(VI) retention. Reasons for this failure, which has been observed by other scientists, are not fully understood. The most likely explanation is that the model is an apparent rather than mechanistic rate law, as previously discussed, because it may not completely account for all reactions and reaction components. Rate coefficients based on batch experiments varied with C_o, which would be expected for pseudo rate coefficients. Unless the concentrations of unaccounted for reaction components remain relatively constant over the course of the experiment, rate coefficients will vary with C_o because they implicitly include concentrations of other reaction components. Much larger changes were observed in Cr(VI) concentrations in column effluent (pulse input) than in the batch solutions. Moreover, in the batch experiment, reaction products are not removed but rather accumulate in the solution and solid phases. In the column experiment, on the other hand, solutes are continually displaced. Also, in the batch experiment, the soil suspension was shaken; but in the column experiment, no such agitation of the solid phase occurred. Thus, reaction rates controlled by transport processes (the usual case in solute retention in soils) rather than by actual chemical reactions will be particularly affected by batch or column flow conditions. Although k_1, k_2, k_3, and k_4 were constant over a limited concentration range in the batch experiment, they did vary over a wide concentration range. Thus, a valid set of rate coefficients from the batch experiment is not readily available to cover the range of concentrations found in the miscible displacement experiment because of the concentration dependence of the rate coefficients. This is particularly true for k_s, which varied systematically with C_o. Criteria for choosing rate coefficients independent of concentration are needed in such cases.

For Olivier soil, predicted BTCs using the SOTS model indicate that the use of batch rate coefficients from either C_o of 10 or 25 mg L^{-1} provided surprisingly good overall descriptions of the experimental results (Figure 5-15). Less than adequate predictions were obtained for the highly kinetic Cecil and Windsor soils, however. In fact, no one set of batch rate coefficients successfully described both the adsorption and the desorption sides of Windsor or Cecil BTCs. For both soils, closest predictions were realized using batch rate coefficients from C_o values of 10 or 25 mg L^{-1}. This is a similar finding to that based on the predictions for Olivier soil.

6

MOBILE–IMMOBILE OR TWO-REGION TRANSPORT APPROACHES

A number of experimental studies have demonstrated early breakthrough results and tailing with nonsymmetrical concentration distributions of effluent break-through results (BTCs). Discrepancies from symmetrical or ideal behavior for several solutes led to the concept of solute transfer between mobile and immobile waters. It was postulated that tailing under unsaturated conditions was perhaps due to the fact that larger pores are eliminated for transport and the proportion of the water that does not readily move within the soil increased. This fraction of water was referred to as stagnant or immobile water. A decrease in water content increases the fraction of air-filled macropores, resulting in the creation of additional dead endpores that depend on diffusion processes to attain equilibrium with a displacing solution. However, the conceptual approach of mobile–immobile or two-region behavior is perhaps more intuitively applicable for well-structured or aggregated soils under either saturated or unsaturated flow. Here, one may assume that within soil aggregates, where micropores are dominant, diffusion is the primary process. In contrast, convection and dispersion are the dominant processes in the macro (or intra-aggregate) pore spaces that occur between large aggregates or structural units (Coats and Smith, 1964; van Genuchten and Wierenga, 1976).

GENERAL FORMULATION

The equations describing the movement for a nonreactive solute through a porous media having mobile and immobile water fractions are

$$\Theta^m \frac{\partial C^m}{\partial t} = \Theta^m D^m \frac{\partial^2 C^m}{\partial x^2} - v^m \Theta^m \frac{\partial C^m}{\partial x} - \alpha(C^m - C^{im}) \qquad (6\text{-}1)$$

and:

$$\Theta^{im} \frac{\partial C^{im}}{\partial t} = \alpha(C^m - C^{im}) \qquad (6\text{-}2)$$

135

Equation 6-1 is a modified version of the convection–dispersion equation where Θ^m and Θ^{im} are mobile and immobile water fractions (cm^3 cm^{-3}), respectively. The terms C^m and C^{im} are the concentrations in the mobile and immobile water (μg cm^{-3}), x is depth (cm), and t is time (h). In addition, D^m (cm^2 h^{-1}) and v^m (cm h^{-1}) are the hydrodynamic dispersion coefficient and the average pore-water velocity in the mobile region, respectively. Also, it is also assumed that the immobile water (Θ^{im}) is located inside aggregate pores (inter-aggregate) where solute transfer occurs by diffusion only. In eq. 6-2, α is a mass transfer coefficient (h^{-1}) that governs the transfer of solutes between the mobile- and immobile-water phases in an analogous manner to a diffusion process.

The mobile–immobile concept represented by eqs. 6-1 and 6-2 may be generalized for the transport of reactive solutes. Incorporation of reversible and irreversible retention for reactive solutes in eqs. 6-1 and 6-2 yields:

$$\Theta^m \frac{\partial C^m}{\partial t} + f\rho \frac{\partial S^m}{\partial t} = \Theta^m D \frac{\partial^2 C^m}{\partial x^2} - v^m \Theta^m \frac{\partial C^m}{\partial x} - \alpha(C^m - C^{im}) - Q^m \quad (6\text{-}3)$$

and

$$\Theta^{im} \frac{\partial C^{im}}{\partial t} + (1 - f)\rho \frac{\partial S^{im}}{\partial t} = \alpha(C^m - C^{im}) - Q^{im} \quad (6\text{-}4)$$

where ρ is soil bulk density (g cm^{-3}). Here the soil matrix is divided into two regions (or sites) where a fraction f is a dynamic or easily accessible region and the remaining fraction is a stagnant or less accessible region (see Figure 6-1).

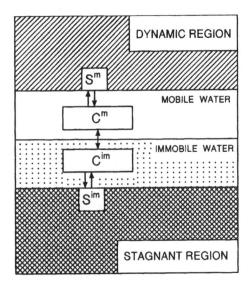

Figure 6-1 Schematic diagram of the mobile–immobile (two-region) concept.

The dynamic region is located close to the mobile phase, whereas the stagnant region is in contact with the immobile phase. Moreover, S^m and S^{im} are the amounts of solutes sorbed in the dynamic and stagnant regions (μg per gram soil), respectively. Also, Q^m and Q^{im} are sink (or source) terms associated with the mobile and immobile water regions, respectively. Therefore, Q^m and Q^{im} represent rates of irreversible type reactions. These terms must be distinguished from S^m and S^{im}, which represent reversible sorbed solutes in the dynamic and stagnant regions, respectively.

The mobile–immobile approach has been successful in describing the fate of several pesticides in soils when linear and Freundlich reversible reactions were considered (van Genuchten et al., 1977). However, it is often necessary to include kinetic rather than equilibrium reactions to account for the degradation of pesticides in soils (Rao et al., 1979). This model was expanded to three-dimensional flow by Goltz and Roberts (1986, 1988). Analytical solutions to equations 6-3 and 6-4 under different initial and boundary conditions can be found in van Genuchten and Wierenga (1976) and De Smedt and Wierenga (1979). Toride et al. (1993) provided a set of analytical solutions for the two-region model with first-order degradation and zero-order production for semi-infinite soil systems. The mobile–immobile model was also analyzed with the method of moments by Sardin et al. (1993) and Valocchi (1990). The mobile–immobile concept was capable of explaining the early breakthrough and extensive tailing of tritium BTCs in porous media (van Genuchten and Wierenga, 1977). Yasuda et al. (1994) found that the mobile–immobile approach best described Br BTCs at different soil depths. van Genuchten et al. (1977) also found that the mobile–immobile approach provided better description of 2,4,5-T BTC in soils when the local (chemical) equilibrium assumption was used.

The mobile–immobile approach has been successfully used to describe heavy metal transport in soils when adsorption was considered as a Langmuir kinetic along with a first-order irreversible reaction (Selim and Amacher, 1988). The mobile–immobile approach received only limited success when extended to describe the transport and exchange of ions in soils for binary (Ca-Mg) and ternary (Ca-Mg-Na) systems (for a review see Selim, 1992). Li and Ghodrati (1994, 1995) compared the mobile–immobile model with the classical CDE and a stochastic model in soils with earthworm holes or root channels, and concluded that, although mobile–immobile model provided best fit to their experimental data, the two-region model does not describe the complexity of solute behavior in porous systems.

The mobile–immobile concept is commonly referred to as the two-region model and is regarded as a mechanistic approach where physical nonequilibrium is a controlling mechanism of solute behavior in porous media. On the other hand, chemically controlled heterogeneous reactions are the governing mechanisms for the two-site (equilibrium/kinetic) approaches (see Chapter 5). However, one can show that the two models are analogous mathematically. Therefore, analyses of data sets of effluent results from miscible displacement experiments alone are not sufficient to differentiate between physical and chemical processes as causes for often observed apparent nonequilibrium behavior in soils. The similarity of the two transport models also means that the two formulations can be used in macroscopic and semi-empirical manners without having to delineate the exact physical and chemical processes on the microscopic level (Nkedi-Kizza et al., 1984; Selim and Amacher, 1988).

ESTIMATION OF α

One major assumption of the mobile–immobile model is that of uniform solute distribution in each water phase. In addition, solute transfer between the two water phases is assumed to follow an empirical first-order diffusion. An alternative to this approach is to assume spherical aggregate geometry with water within the aggregates as the immobile water phase where solute distribution in the sphere is not considered uniform (Rao et al., 1980a). Moreover, solute diffusion into the aggregates can be governed by Fick's second law, which may be expressed in spherical coordinates as:

$$\frac{\partial C^{im}(r,t)}{\partial t} = D_e\left\{\frac{\partial^2 C^{im}}{\partial r^2} + \frac{2}{r}\frac{\partial C^{im}}{\partial r}\right\} \tag{6-5}$$

where $D_e (= D_o\tau^2)$ is an effective molecular diffusion coefficient, D_o is molecular diffusion in water, τ^2 is a tortuosity factor (< 1), and r is the radial coordinate in a sphere of diameter a. Here, D_e was assumed to be independent of concentration within the aggregate C^{im}. Average concentration in the sphere can be calculated using:

$$\bar{C}^{im}(t) = \frac{3}{a}\int_0^a r^3 C^{im}(r,t)\,dr \tag{6-6}$$

As a result, Rao et al. (1980a,b) derived an approximate expression for α assuming spherical aggregates as:

$$\alpha = \left[\frac{D_e\Theta^{im}}{a^2}\right]\alpha^*(t) \tag{6-7}$$

where the parameter α^* is estimated based on the aggregate size a, D_o, t, and F the fraction of the mobile to total water content ($F = \Theta^m/\Theta$) (see Rao et al., 1980a). As a result, this α or $\alpha(t)$ in eq. 6-7 is time-dependent and approximates the diffusion process in a sphere. In an earlier attempt to arrive at an expression for an overall dispersion D for nonreactive solute transport in spherical aggregates, Passioura (1971) approximated the overall D for soils composed of spherical aggregates as:

$$D = D^m F + \frac{(1 - F)a^2 v^2}{15 D_e} \tag{6-8}$$

with the constraint:

$$\frac{(1 - F)D_e L}{a^2 v^2} > 0.3 \tag{6-9}$$

where L is solute transport length. As evidenced by eq. 6-8, when physical nonequilibrium is dominant, the overall D increases with increasing velocity and aggregate sizes. Equation 6-8 was extended to include a reactive solute with a retardation factor R (van Genuchten and Dalton, 1986; Parker and Valocchi, 1986):

$$D = D^m F + \frac{(1 - F)a^2 v^2 [R^{im}]^2}{15 D_e R^2} \tag{6-10}$$

where R^{im} is the retardation factor associated with the immobile phase, R is an overall retardation factor ($\Theta R = \Theta^{im}R^{im} + \Theta^m R^m$). Similar effective dispersion coefficients were obtained for rectangular aggregates with half width of a_l as:

$$D = D^m F + \frac{(1 - F)a_l^2 v^2 [R^{im}]^2}{3 D_e R^2} \tag{6-11}$$

An overall D can also be derived from empirical first-order mass transfer where uniform solute distribution in the immobile water phase may be assumed. This was carried out by De Smedt and Wierenga (1984) who derived an expression for D for nonreactive solutes for long columns as:

$$D = D^m F + \frac{\Theta(1 - F)^2 v^2}{\alpha} \tag{6-12}$$

and for reactive solutes (van Genuchten and Dalton, 1986) as:

$$D = D^m F + \frac{\Theta(1 - F)^2 v^2 [R^{im}]^2}{\alpha R^2} \tag{6-13}$$

Comparing eqs. 6-8 and 6-12, an equivalent first-order transfer coefficient (α) for spherical aggregates is thus obtained:

$$\alpha = \frac{15 D_e (1 - F)\Theta}{a^2} \tag{6-14}$$

This equation can also be obtained using moment analysis (Valocchi, 1990) and has been used in solute transport (Selim et al., 1987; Selim and Amacher, 1988). Similar α expressions were obtained for a rectangular aggregate as:

$$\alpha = \frac{3D_e(1 - F)\Theta}{a_l^2} \tag{6-15}$$

As a result, a more general formulation for estimating α is

$$\alpha = \frac{n(1 - F)\Theta D_e}{a_e^2} \tag{6-16}$$

where n is a geometry factor and a_e is an average effective diffusion length. Details can be found in van Genuchten and Dalton (1986) and van Genuchten (1985). Equation 6-14 has been used to estimate α in modeling solute transport in porous media (Selim and Gaston, 1990; Goltz and Roberts, 1988).

ESTIMATION OF Θ^m AND Θ^{im}

A common way of estimating the mobile and immobile water contents is by use of curve-fitting of tracer breakthrough results (Li and Ghodrati, 1994; van Genuchten and Wierenga, 1977). De Smedt and Wierenga (1984) found $\Theta_m = 0.853\Theta$ is applicable for unsaturated glass beads with diameters in the neighborhood of 100 μm. Alternatively, a direct method of estimating Θ^m and Θ^{im} is by measuring the soil water content at some arbitrary water tension (ψ). Smettem and Kirkby (1990) used water content at $\psi = 14$ cm as the matching point between the interaggregate (macro-) and the intraaggregate (micro-) porosity by examining the ψ-Θ soil moisture characteristic curve. Jarvis et al. (1991) estimated macroporosity from specific yield under water tension of 100 cm. Other water tensions used to differentiate macropores from micropores are 3 cm (Luxmoore, 1981), 10 cm (Wilson et al., 1992), 20 cm (Selim et al., 1987), and 80 cm (Nkedi-Kizza et al., 1982). A list of water tensions used by different authors was provided by Chen and Wagenet (1992). The equivalent diameters at these water tensions range from 10 to 10,000 μm.

Another experimental measurement of Θ^m is based on the following mass balance equation:

$$\Theta C = \Theta^m C^m + \Theta^{im} C^{im} \tag{6-17}$$

When α is small enough to assume $C^{im} = 0$ and $C^m = C_o$ (input concentration) at certain infiltration time t, the approximate equation is obtained:

$$\Theta^m = \Theta \frac{C}{C^m} = \Theta \frac{C}{C_o} \tag{6-18}$$

Applications of this method can be found in Clothier et al. (1992) and Jaynes et al. (1995). By assuming that a tracer concentration in the mobile water phase (C^m) equals input solution concentration (C_o), Jaynes et al. (1995) derived the following formula from equations (6-2) and (6-18):

$$\ln\left(1 - \frac{C}{C_0}\right) = -\frac{\alpha t}{\Theta^{im}} + \ln\left(\frac{\Theta^{im}}{\Theta}\right) \tag{6-19}$$

α and Θ^{im} can be estimated by plotting $\ln(1 - C/C_o)$ versus application (infiltration) time. However, the assumption of $C^m = C_o$ associated with this method is questionable and may not be correct as long as $\alpha \neq 0$. A slightly different approach was used by Goltz and Roberts (1988) to estimate the fraction of mobile water as the ratio of velocity calculated from hydraulic conductivity to the velocity measured from tracer experiment.

A SECOND-ORDER APPROACH

In the previous chapter, the second-order reactions associated with sites 1 and 2 were considered as kinetically controlled, heterogeneous chemical retention reactions (Rubin, 1983). One can assume that these processes are predominantly controlled by surface reactions of adsorption and exchange. In this sense, the second-order model is along the same lines as the earlier two-site model of Selim et al. (1976) and Cameron and Klute (1977). Another type of two-site model is that of Villermaux (1974), which is capable of describing breakthrough curves (BTCs) from chromatography columns having two concentration maxima.

In this section, the second-order concept is invoked where processes of solute retention were controlled by two types of reactions; namely, chemically controlled heterogeneous reaction and the other physically controlled reaction (Rubin, 1983). The chemically controlled heterogeneous reaction was considered to be governed according to the second-order approach. In the meantime, the physically controlled reaction is chosen to be described by diffusion or mass transfer based on the mobile–immobile concept (Coats and Smith, 1964; van Genuchten and Wierenga, 1976). A comparison between the mobile–immobile concept and that of the two-site approach indicates that the dynamic and stagnant regions for solute retention are analogous to sites 1 and 2 of the two-site concept. Nkedi-Kizza et al. (1984) presented a detailed discussion on the equivalence of the mobile–immobile and the equilibrium–kinetic two-site models.

An important feature of the second-order approach is that an adsorption maximum (or capacity) is assumed. This maximum represents the total number adsorption sites per unit mass or volume of the soil matrix. It is also considered an intrinsic property of an individual soil and is thus assumed constant. In previous two-region models, a finite amount of sites has not been specified and thus an adsorption maximum is never attained (e.g., van Genuchten and Wierenga, 1976; Rao et al., 1979; Brusseau et al., 1989; van Genuchten and Wagenet, 1989). Specifically, in eqs. 6-3 and 6-4, the dimensionless term f denotes the ratio or fraction of dynamic or active sites to the maximum or total adsorption sites S_{max} (μg g^{-1} soil). In addition, the terms $(\partial S^m/\partial t)$ and $(\partial S^{im}/\partial t)$ represent the rates of reversible heavy metal reactions between C in soil solution and that present on matrix surfaces in the mobile (or dynamic) and the immobile regions, respectively. Moreover, irreversible reaction of heavy metals was incorporated in this model, as may be seen by the inclusion of the sink terms Q^m and Q^{im} (μg cm^{-3} h^{-1}) in eqs. 6-3 and 6-4, respectively. It is assumed that irreversible retention or removal from solution will occur separately in the mobile and immobile water regions. However, the governing mechanism of retention for each region was assumed to follow first-order type reactions. Specifically, it is assumed that irreversible retention for the mobile and immobile regions be considered in this transport model as follows:

$$Q^m = \Theta^m k_s C^m \tag{6-20}$$

$$Q^{im} = \Theta^{im} k_s C^{im} \tag{6-21}$$

where k_s is an irreversible rate coefficient (h^{-1}) common to both regions. These formulations for the sink terms do not occur elsewhere in the literature and were first proposed by Selim and Amacher (1988). Since soil matrix surfaces may behave in a separate manner to heavy metal retention, it is conceivable that the rate of irreversible reactions in the mobile region are characteristically different from that for the mobile region. One way to achieve this is to distinguish between the rate coefficient (k_s) controlling the reaction for the two regions; e.g., k_s^m and k_s^{im} in eqs. 6-20 and 6-21, respectively. However, such a distinction in reaction coefficients was not incorporated into this model and a common parameter k_s was thus used.

The retention mechanism associated with the mobile and immobile phases of eqs. 6-3 and 6-4 was considered as an equilibrium linear sorption by van Genuchten and Wierenga (1976) and was extended to the nonlinear or Freundlich type by Rao et al. (1979). Multiple ion retention expressed on the basis of ion-exchange equilibrium reactions was successfully incorporated into the mobile–immobile model by van Eijkeren and Loch (1984) and Selim et al. (1987). Here, it is considered that reversible solute reactions be governed by the second-order kinetic approach. Specifically, the rates of reaction for S^m and S^{im} were considered as (Selim and Amacher, 1988):

$$\rho \frac{\partial S^m}{\partial t} = \Theta^m k_1 \phi^m C^m - \rho k_2 S^m \tag{6-22}$$

and

$$\rho \frac{\partial S^{im}}{\partial t} = \Theta^{im} k_1 \phi^{im} C^{im} - \rho k_2 S^{im} \tag{6-23}$$

where k_1 and k_2 are forward and backward rate coefficients (h^{-1}), respectively. Here ϕ^m and ϕ^{im} represent the vacant or unfilled sites (μg per gram soil) within the dynamic and the stagnant regions, respectively. In addition, the terms ϕ^m and ϕ^{im} can be expressed as:

$$\phi^m = S^m_{max} - S^m = f S_{max} - S^m \tag{6-24}$$

$$\phi^{im} = S^{im}_{max} - S^{im} = (1 - f) S_{max} - S^{im} \tag{6-25}$$

where S_{max}, S^m_{max} and S^{im}_{max} are the total amount of the sites in the soil matrix, total sites in the dynamic region, and the total in the less accessible region (mg kg^{-1}), respectively. These terms are related by:

$$S_{max} = S^m_{max} + S^{im}_{max} \tag{6-26}$$

We also assume S_{max} to represent the combined total of occupied and unoccupied sites, i.e., maximum adsorption capacity of an individual soil, and is regarded as an intrinsic property of the soil.

An important feature of the second-order retention approach (eqs. 6-22 and 6-23) is that similar reaction rate coefficients (k_1 and k_2) associated with the dynamic and stagnant regions were chosen. Specifically, it is assumed that the retention mechanism is equally valid for the two regions of the porous media. A similar assumption was made by van Genuchten and Wierenga (1976) for equilibrium linear and Freundlich-type reactions and by Selim et al. (1987) for selectivity coefficients for homovalent ion-exchange reactions. Specifically, as $t \to \infty$, i.e., when both the dynamic (or active) sites and the sites in the stagnant region achieve local equilibrium, eqs. 6-22 and 6-23 yield the following expressions. For the active sites associated with the mobile region:

$$\Theta^m k_1 \phi^m C^m - \rho k_2 S^m = 0 \quad \text{as} \quad t \to \infty \tag{6-27}$$

or

$$\frac{S^m}{\phi^m C^m} - \frac{\Theta^m k_1}{\rho k_2} = \omega_1 \quad as \quad t \to \infty \tag{6-28}$$

and for the sites associated with the immobile region, we have

$$\Theta^{im} k_1 \phi^{im} C^{im} - \rho k_2 S^{im} = 0 \quad \text{as} \quad t \to \infty \tag{6-29}$$

or

$$\frac{S^{im}}{\phi^{im}C^{im}} = \frac{\Theta^{im}k_1}{\rho k_2} = \omega_2 \quad as \quad t \to \infty \tag{6-30}$$

Here, ω_1 and ω_2 represent equilibrium constants for the retention reactions associated with the mobile and immobile regions, respectively. The formulation of eqs. 6-28 and 6-30 are analogous to expressions derived for the second-order two-site model discussed previously. In this sense, the equilibrium constants ω_1 and ω_2 resemble the Langmuir coefficients with S_{max} as the maximum sorption capacity (Selim and Amacher, 1988). These equilibrium constants are also analogous to the selectivity coefficients associated with ion exchange reactions (see Selim et al., 1987).

The dimensionless form of eqs. 6-3, 6-4, 6-22, and 6-23 are

$$\Omega f \frac{\partial s^m}{\partial T} + F \frac{\partial c^m}{\partial T} = \frac{F}{P} \frac{\partial^2 c^m}{\partial X^2} - \frac{\partial c^m}{\partial X} - \bar{\alpha}(1 - F)(c^m - c^{im}) - Fk_s c^m \tag{6-31}$$

$$\frac{\partial c^{im}}{\partial T} + \Omega \frac{1 - f}{1 - F} \frac{\partial s^{im}}{\partial T} = \bar{\alpha}(c^m - c^{im}) - k_s c^{im} \tag{6-32}$$

$$\frac{\partial s^m}{\partial T} = Fk_1 \Phi^m c^m - k_2 s^m \tag{6-33}$$

$$\frac{\partial s^{im}}{\partial T} = (1 - F)' k_1 \Phi^{im} c^{im} - k_2 s^{im} \tag{6-34}$$

where

$$c^m = \frac{C^m}{C_o}, \qquad c^{im} = \frac{C^{im}}{C_o} \tag{6-35}$$

$$s^m = S^m/S_{max}, \quad s^{im} = S^{im}/S_{max}, \quad \Phi^m = \phi^m/S_{max}, \quad \Phi^m = \phi^m/S_{max} \tag{6-36}$$

$$P = qL/D^m \Theta \tag{6-37}$$

$$\bar{\alpha} = \alpha L/q(1 - F) \tag{6-38}$$

$$F = \Theta^m/\Theta \tag{6-39}$$

$$X = x/L \tag{6-40}$$

$$T = qt/L\Theta \tag{6-41}$$

where T dimensionless time equivalent to the number of pore volumes leached through a soil column of length L, and P is the Peclet number Brenner (1962). In addition, we have defined:

$$\Omega = S_{max}\rho/C_o\Theta \tag{6-42}$$

$$\kappa_s = k_s\Theta L/q \tag{6-43}$$

$$\kappa_1 = k_1\Theta^2 C_o L/\rho q \tag{6-44}$$

$$\kappa_2 = k_2\Theta L/q \tag{6-45}$$

Here, κ_s, κ_1, and κ_2 are dimensionless kinetic rate coefficients that incorporate q and L.

Initial and Boundary Conditions

The corresponding initial and boundary conditions associated with the second-order mobile–immobile model can be expressed as:

$$C^m = C^{im} = C_i \qquad\qquad (t = 0, 0 < x < L) \tag{6-46}$$
$$S^m = S^{im} = S_i \qquad\qquad (t = 0, 0 < x < L) \tag{6-47}$$
$$vC_o = qC^m - \Theta^m D^m \partial C^m/\partial x \qquad (x = 0, t < t_p) \tag{6-48}$$
$$0 = qC^m - \Theta^m D^m \partial C^m/\partial x \qquad (x = 0, t > t_p) \tag{6-49}$$
$$\partial C^m/\partial x = 0 \qquad\qquad (x = L, t \geq 0). \tag{6-50}$$

These conditions are similar to those described earlier for the transport of a solute pulse (input) in a uniform soil having a finite length, L, where a steady water flux q was maintained constant. The soil column is considered as having uniform retention properties as well as having uniform ρ and Θ. It is further assumed that equilibrium conditions exist between the solute present in the soil solution of the mobile water phase (i.e., interaggregate) and that present in the immobile (or inter-aggregate) phase. This necessary condition is expressed by eqs. 6-46 and 6-47. Uniform initial conditions were assumed along the soil column. It is assumed that an input heavy metal solution pulse having a concentration C_o was applied at the soil surface for a time duration t_p and was then followed by a solute-free solution. As a result, at the soil surface, the third-type boundary conditions were those of eqs. 6-48 and 6-49. In a dimensionless form, the boundary conditions can be expressed as:

$$1 = c^m - (1/P)\partial c^m/\partial X, \qquad (X = 0, T < T_p) \tag{6-51}$$

$$0 = c^m - (1/P)\partial c^m/\partial X, \qquad (X = 0, T > T_p) \tag{6-52}$$

and at $x = L$, we have

$$\partial c^m/\partial X = 0, \qquad (X = 1, T > 0) \tag{6-53}$$

where T_p is dimensionless time of input pulse duration of the applied solute and represents the amount of applied pore volumes of input solution.

SENSITIVITY ANALYSIS

Figures 6-2 through 6-6 are examples of simulated breakthrough curves (BTCs) to illustrate the sensitivity of the proposed second-order reaction, when incorporated into the mobile–immobile concept, to various model parameters. As shown, several features of the mobile–immobile concept dominate the behavior of solute transport and thus the shape of simulated BTCs. For this reason, we restrict the discussion here to the influence of parameters pertaining to the proposed second-order mechanism. Specifically, the influence of κ_1 and κ_2, and ω on solute retention were examined. Other parameters such as D^m and q have been rigorously examined in earlier studies by Coats and Smith (1964) and van Genuchten and Wierenga (1976).

For the simulations shown in Figures 6-2 to 6-6, initial conditions, volume of input pulse, and model parameters were identical to those used previously for the second-order, two-site model, where $C_i = S_i = 0$ within the mobile and immobile regions. Specifically, the parameters chosen were: $L = 10$ cm, $D^m = 1$ cm^2 h^{-1}, $\rho = 1.2$ g cm^{-3}, $f = 0.50$, $\Theta = 0.40$ cm^3 cm^{-3}, $F = \Theta^m/\Theta = 0.5$, $C_o = 100$ mg L^{-1}, and a Peclet number $P = 25$. Moreover, unless otherwise stated, the values selected for the dimensionless parameters κ_1, κ_2, κ_s, Ω, and $\overline{\alpha}$ used were 1, 1, 0, 5, and 1, respectively. It is assumed a solute pulse was applied to a fully water-saturated soil column initially devoid of a particular heavy metal of interest. In addition, a steady water flow velocity (q) was maintained constant with a Peclet number P of 25. The length of the pulse was assumed to be 3 pore volumes, which was then followed by several pore volumes of metal-free solution.

The influence of the reaction rate coefficients on the shape of the BTCs is illustrated in Figure 6-2. Here the values of κ_1 and κ_2 were varied simultaneously provided that κ_1/κ_2 (and ω_1 and ω_2) remained invariant. For the nonreactive case

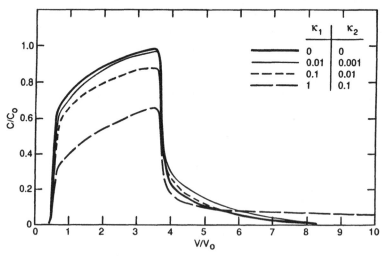

Figure 6-2 Effluent concentration distributions for different values of κ_1 and κ_2 the SOMIM model.

($\kappa_1 = \kappa_2 = 0$), the highest effluent peak concentration and least tailing were observed. As the rate of reactions increased simultaneously, solute peak concentrations decreased and excessive tailing of the BTCs was observed. However, the arrival time or the location of peak concentration was not influenced by increasing the rates of reactions.

The effect of increasing values of the equilibrium constant ω, which represents the ratio κ_1/κ_2, on the shape of BTCs is shown in Figure 6-3. Here, a constant value for κ_2 of 1 was chosen, whereas κ_1 was allowed to vary. For all BTCs shown in Figure 6-3, the values of ω_1 and ω_2 were equal (since $F = 0.5$). As a result, we refer to simply ω rather than ω_1 and ω_2. The results indicate that as the forward rate of reaction (κ_1) increased, an increase in solute retardation or a right shift of the BTCs was observed. This shift of the BTCs was accompanied by an increase in solute retention (i.e., a decrease of the amount of solute in the effluent, based on the area under the curve) and a lowering of peak concentrations. Similar behavior was observed for the influence of the dimensionless transfer coefficient ($\overline{\alpha}$) on the shape of the BTCs, as may be seen from the BTCs of Figure 6-4. For exceedingly large values of $\overline{\alpha}$ (>2), the diffusion between the mobile and immobile phases became more rapid. Therefore, equilibrium conditions between the two phases is nearly attained (Valocchi, 1985).

Figure 6-5 shows breakthrough curves (BTCs) of a reactive solute for several values of Ω. The figure indicates that the shape of the BTCs is influenced drastically by the value of Ω. This is largely due to the nonlinearity of the proposed second-order retention mechanism. As given by eq. 6-25, Ω represents the ratio of total sites (S_{max}) to input (pulse) solute concentration (C_o). Therefore, for small Ω values (e.g., $\Omega = 0.1$), the simulated BTC is very similar to that for a nonretarded solute due to the limited amount of sites (S_{max}) in comparison to

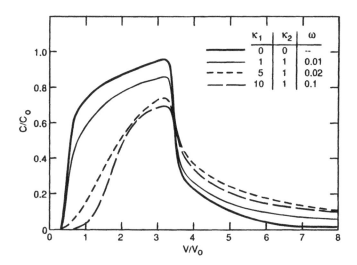

Figure 6-3 Effluent concentration distributions for different values of the parameter ω using the SOMIM model. Values of κ_1 and κ_2 are also shown.

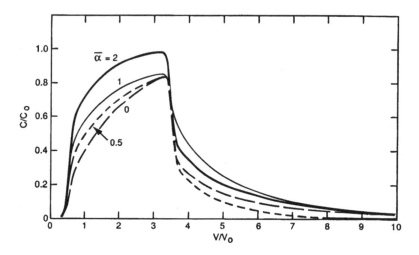

Figure 6-4 Effluent concentration distributions for different values of the dimensionless mass transfer parameter $(\overline{\alpha})$ of the SOMIM model.

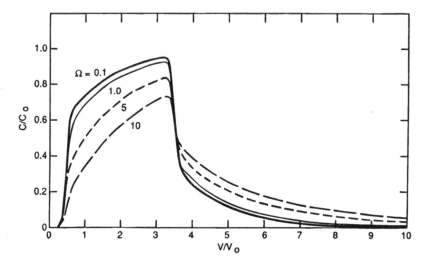

Figure 6-5 Effluent concentration distributions for different values of the parameter Ω of the SOMIM model.

C_o. In contrast, large values of Ω resulted in BTCs which indicate increased retention, as manifested by the right shift of peak concentration of the BTCs. In addition, for high Ω values, extensive tailing as well as an overall decrease of effluent concentration were observed. The influence of the parameter f, which represents the fraction of active or dynamic sites within the mobile region to the total amount of sites on the behavior of solute retention and transport, is shown in Figure 6-6 for several values of f. There are similar features between these BTCs and those illustrated in the previous figures. For $f = 1$, all the sites are

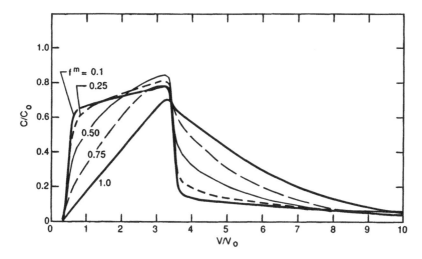

Figure 6-6 Effluent concentration distributions for different values of the fraction of active sites f of the SOMIM model.

active sites and thus there is no solute retention by the sites present within the immobile region (i.e., stagnant sites). As the contribution of the stagnant sites increases (or f decreases), the shape of the BTCs becomes increasingly less kinetic with significant increases of the tailing of the desorption side of the BTCs.

In the BTCs shown in Figures 6-2 through 6-6, the irreversible retention mechanism for heavy metal removal (via the sink term) was ignored. The influence of the irreversible kinetic reaction (e.g., precipitation, etc.) is a straight-forward one and is thus not shown. This is manifested by the lowering of solute concentration for the overall BTC for increasing values of k_s. Since a first-order reaction was assumed, the lowering of the BTC is proportional to the solution concentration. The influence of other parameters on the behavior of solute in soils with the second-order mobile–immobile model such as P, D^m, and q have been studied elsewhere (van Genuchten and Wierenga, 1976).

APPLICATIONS

The capability of the second-order mobile–immobile model to describe the transport of heavy metals in soils was examined for hexavalent chromium [Cr(VI)] by Selim and Amacher (1988). Their Cr(VI) miscible displacement results and model predictions for three soils are shown in Figures 6-7 to 6-9. To obtain the predictions, several assumptions were necessary for the estimation of model parameters. The sorption maximum (S_{max}) was estimated from kinetic adsorption isotherms. In addition, the ratio of the mobile to total water contents (Θ^m/Θ) was estimated based on soil–moisture retention relations for each soil. Selim and Amacher (1988) also assumed that the fraction of sites f is the same as the relative amount of water in the two regions, i.e., $f = F (= \Theta^m/\Theta)$. Such an assumption

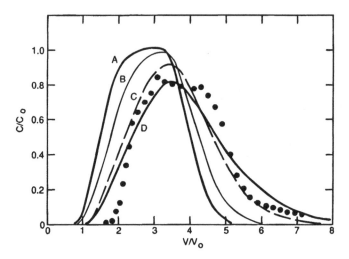

Figure 6-7 Effluent concentration distributions for Cr (VI) in Olivier soil. Curves A, B, C, and D are predictions using the second-order mobile–immobile model with batch rate coefficients for C_o of 25, 10, 5, and 1 mg/L, respectively.

was made because independent measurement of f was not available (van Genuchten and Dalton, 1986). Selim et al. (1987) successfully used such estimates of f for a well-aggregated soil. Estimates for α were obtained using eq. 6-14, with average aggregate sizes of 0.01, 0.01, and 0.005 cm for the Windsor, Olivier, and Cecil soils used, respectively. In addition, D_o for Cr(VI) diffusion was assumed to be 10^{-9} cm^2 sec^{-2} for all three soils. Barber (1984) compiled diffusion

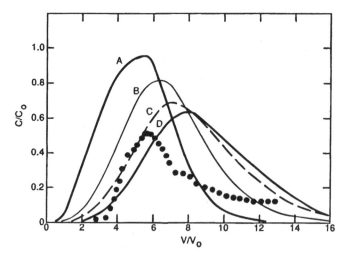

Figure 6-8 Effluent concentration distributions for Cr(VI) in Windsor soil. Curves A, B, C, and D are predictions using the second-order mobile–immobile model with batch rate coefficients for C_o of 25, 5, 2, and 1 mg/L, respectively.

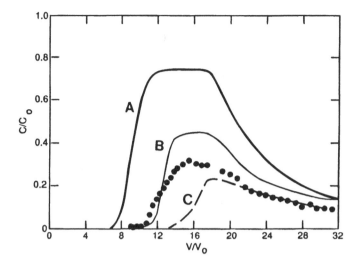

Figure 6-9 Effluent concentration distributions for Cr(VI) in Cecil soil. Curves A, B, and C are predictions using the second-order mobile–immobile model with batch rate coefficients for C_o of 25, 10, and 5 mg/L, respectively.

coefficients for a number of ions in soils with values for phosphate ($H_2PO_4^-$) ranging from 10^{-8} to 10^{-11} cm^2 sec^{-1} (for water, $D_o = 10^{-6}$ cm^2 sec^{-1}). Since chromate and phosphate have somewhat similar behaviors in soils, one can assume that diffusion coefficients for chromate would be equivalent to those for phosphate. Such values were used to estimate D^m using the formulation of eq. 6-8. Selim and Amacher (1988) stated that attempts to utilize tracer data sets from tritium and chloride-36 for parameter estimation of α and D^m were not successful. The values obtained using least-squares parameter optimization were inconsistent and ill-defined due to large parameter standard errors and were often physically unrealistic (e.g., a retardation factor R >> 1). Perhaps these results are due to local equilibrium conditions between the mobile and immobile regions for the two tracers (Rubin, 1983; Parker and Valocchi, 1986).

Values for the rate coefficients used in the second-order, mobile–immobile model were those calculated from batch kinetic results. Selim and Amacher (1988) used k_1, k_2, and k_s values from a three-parameter version of the second-order model described earlier. Predicted BTCs were obtained using different sets of batch rate coefficients due to their strong dependence on input concentrations (C_o). Closest predictions to experimental Cr(VI) measurements were obtained from batch rate coefficients at low C_o values ($C_o \leq 10$ mg L^{-1}). Moreover, the use of rate coefficients at higher C_o values resulted in decreased tailing and reduced retardation of the BTCs. These observations are consistent with the second-order, two-site approach discussed in the previous chapter. Overall predictions of measured Cr(VI) using this model may be considered adequate. However, it is conceivable that a set of applicable rate coefficients over the concentration range for Cr(VI) transport experiments cannot be obtained simply by use of the batch procedure described in this study. In addition, several parameters used in model calculations were estimated and not measured, e.g., Θ^m, α, and D. The fraction of active sites f was not estimated; rather, it was assumed equal to the mobile water fraction (Θ^m/Θ). Amacher and Selim (1988) argued that it is likely that improved model predictions could be obtained if such parameters could be measured independently. They also postulated that other possible factors responsible for these predictions may be due in part to the lack of nonequilibrium conditions between the mobile and immobile fractions (Valocchi, 1985; Parker and Valocchi, 1986).

A MODIFIED TWO-REGION APPROACH

Due to the uncertainty of obtaining an independent measurement for the fraction of sites f, Selim and Ma (1995) reexamined the original assumption of the mobile–immobile approach within the scope of the second-order formulation described above. In the modified approach, they considered the dynamic and stagnant soil regions in the soil as a continuum, and connected to one another (Figure 6-10). Solute retention may occur concurrently in the dynamic and stagnant regions until equilibrium conditions are attained or all vacant sites for a soil aggregate become occupied (filled). They proposed that the rates of retention reactions in the mobile and immobile phases are a function of the total vacant sites in the soil. Specifically, this modified approach does not distinguish between the fraction of sites associated with the dynamic region from that of the stagnant region. That is, the amount retained from the mobile phase, for example, affects the total amount of vacant sites for retention of solutes in the immobile water phase, and vice versa. In fact, the fraction of sites f has been shown to be highly

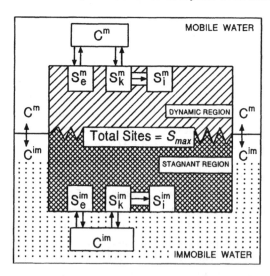

Figure 6-10 Schematic diagram of a modified mobile–immobile (two-region) concept.

affected by experimental conditions such as particle size, water flux, solute concentration, and species considered (van Genuchten and Wierenga, 1977; Nkedi-Kizza et al., 1983). Selim and Ma (1995) also assumed that the second-order approach accounts for two reversible kinetic reactions and one irreversible reaction. Specifically, S_e and S_k are associated with reversible and S_i with irreversible reactions. According to the second-order rate law, the rate of reaction is not only a function of solute concentration in solution, but also of the amount of available retention sites on matrix surfaces. As the sites become filled or occupied by the retained solute, the amount of vacant or unfilled sites that we denote as ϕ (μg per gram soil) approaches zero. In the meantime, the amount of solutes retained by the soil matrix (S) approaches the total capacity or maximum sorption sites S_{max}.

Incorporating the modified concept to the mobile–immobile approach, the transport CD equation with reactions in the dynamic soil region can be rewritten as:

$$\Theta^m \frac{\partial C^m}{\partial t} + \rho \frac{\partial S^m}{\partial t} = \Theta^m D^m \frac{\partial^2 C^m}{\partial x^2} - v^m \Theta^m \frac{\partial C^m}{\partial x} - \alpha(C^m - C^{im}) \quad (6\text{-}54)$$

and the associated mass transfer equation as:

$$\Theta^{im} \frac{\partial C^{im}}{\partial t} + \rho \frac{\partial S^{im}}{\partial t} = \alpha(C^m - C^{im}) \quad (6\text{-}55)$$

Let ϕ denotes the total amount of vacant sites. The reactions in the dynamic soil region are

$$S_e^m = K_e \Theta^m C^m \phi \quad (6\text{-}56)$$

$$\frac{\partial S_k^m}{\partial t} = k_1 \Theta^m C^m \phi - k_2 S_k^m - k_s S_k^m \quad (6\text{-}57)$$

$$\frac{\partial S_i^m}{\partial t} = k_s S_k^m \quad (6\text{-}58)$$

For the stagnant region, the reactions are

$$S_e^{im} = K_e \Theta^{im} C^{im} \phi \quad (6\text{-}59)$$

$$\frac{\partial S_k^{im}}{\partial t} = k_1 \Theta^{im} C^{im} \phi - k_2 S_k^{im} - k_s S_k^{im} \quad (6\text{-}60)$$

and

$$\frac{\partial S_i^{im}}{\partial t} = k_s S_k^{im} \tag{6-61}$$

Maximum sorption capacity S_{max} for a given soil is related to the total vacant sites ϕ, according to:

$$S_{max} = \phi + S_e^m + S_k^m + S_e^{im} + S_k^{im} \tag{6-62}$$

At large reaction time when equilibrium is assumed, the following relations express the amount sorbed for the dynamic and stagnant regions, respectively:

$$\frac{S_e^m}{S_{max}} = \left[\frac{K_e \Theta^m C^m}{1 + \omega \Theta \bar{C}}\right], \quad \frac{S_k^m}{S_{max}} = \left[\frac{K_k \Theta^m C^m}{1 + \omega \Theta \bar{C}}\right] \tag{6-63}$$

$$\frac{S_e^{im}}{S_{max}} = \left[\frac{K_e \Theta^{im} C^{im}}{1 + \omega \Theta \bar{C}}\right], \quad \frac{S_k^{im}}{S_{max}} = \left[\frac{K_k \Theta^{im} C^{im}}{1 + \omega \Theta \bar{C}}\right] \tag{6-64}$$

where ω is similar to that developed in the previous model ($\omega = K_c + K_k$) and $\bar{C}$ represents average concentration in solution, ($\Theta \bar{C} = \Theta^m C^{im} + \Theta^{im} C^{im}$). The total amount S can thus be expressed by the following simplified (Langmuir) form:

$$\frac{S}{S_{max}} = \frac{\omega \Theta \bar{C}}{1 + \omega \Theta \bar{C}} \tag{6-65}$$

Therefore, the modified mobile–immobile model approaches a one-site Langmuir isotherm. Moreover, the parameter f is absent in eqs. 6-41 through 6-52 and, at equilibrium, the amounts sorbed in each fraction depend solely on K_e, K_k, and Θ^m. Therefore, the partitioning coefficient f of the two-region concept, which is a difficult parameter to measure, need not be specified and the amounts retained by each soil region is solely a function of the rates of reactions.

Selim and Ma (1995) applied their modified model on miscible displacement results for atrazine. They showed that their modified approach provided good BTC predictions for a wide range of aggregate sizes and flow velocities for a Sharkey clay soil. They also concluded that the modified approach, which requires fewer parameters, is superior to the original model of Selim and Amacher (1988). Application of this approach for predicting the transport of heavy metals in soils has not been made. This modified approach is a promising one since it accounts for retention and transport processes based upon physically as well as chemically heterogeneous reactions.

Although the two-region model concept has been shown to successfully describe the appearance of lack of equilibrium behavior and tailing for a wide range of conditions, this approach has several drawbacks. First, the value of α

is difficult to determine for soils because of the irregular geometric distribution of immobile water pockets; and secondly, the fraction of mobile and immobile water within the system can only be estimated. Thus, two parameters are needed (for nonreacting solute) and they can only be found by curve-fitting of the flow equations to effluent data. Another drawback of the mobile–immobile approach is the inability to identify unique retention reactions associated with the dynamic and stagnant soil regions separately. Due to this difficulty, a general assumption implicitly made is that similar processes and associated parameters occur within both regions. Thus, a common set of model parameters are utilized for both regions. Such an assumption has been made for equilibrium (linear, nonlinear, and ion exchange) as well as kinetic reversible and irreversible reactions. Therefore, this model disregards the heterogeneous nature of various types of sites on matrix surfaces. This is not surprising since soils are not homogeneous systems, but rather are a complex mixture of solids of clay minerals, several oxides/ hydroxides, and organic matter with varying surface properties.

7

MULTICOMPONENT AND COMPETITIVE ION EXCHANGE APPROACHES

A number of scientists have attempted to describe heavy metal transport data in the soil profile under field conditions and in laboratory miscible displacement experiments. Sidle et al. (1977) were among the earliest researchers to utilize the convection–dispersion equation for the description of Cu, Zn, and Cd movement in soils. The retention/release mechanism was reversible and nonlinear (Freundlich equilibrium). Based on experimental results for two soil depths, Sidle et al. (1977) found that model calculations resulted in under-prediction of the mobility of these metals. A similar approach was used by Amoozegar-Fard et al. (1983), where a linear (equilibrium) sorption mechanism was incorporated into the convection–dispersion equation to describe Cr(VI) mobility in soil columns. Swartjes et al. (1992) also utilized a Freundlich sorption mechanism to describe Cd migration in an Orthic Luvisol profile. Cu leaching in soil columns was described based on equilibrium adsorption–desorption, coupled with kinetic solubilization (Montero et al., 1994).

The research mentioned above focused on the description of transport and retention of one heavy metal species only. Such an assumption implies that all other interactions that occur in the soil do not greatly influence the behavior of the heavy metal species under consideration. This simplification is unrealistic and does not represent the soil environment that contains many chemical species having various interactions. One of the early works dealing with the transport and interaction of two species is that of Lai and Jurinak (1972). In their approach, a simple equilibrium reaction was assumed to govern the distribution of two ions between solution and that on soil matrix surfaces. Such an equilibrium reaction was incorporated into the classical convection–dispersion equation for solute transport in porous media. Rubin and James (1973) presented an analysis of multiple ion transport and interactions where equilibrium ion exchange reactions were assumed to govern the competitive ion exchange process. A set of recursion equations were developed to describe the reactions for multiple species of heterovalent ion exchange.

Valocchi et al. (1981) extended the work of Rubin and James (1973) to include multiple ions under conditions of varying ionic strength or total solute concentration in the soil solution. They applied their approach with some success to field results of Ca and Mg displacement in a groundwater aquifer. Selim et al. (1987) and Mansell et al. (1988) incorporated ion exchange reactions with the two-region (mobile–immobile) concept. Such an approach was generally successful in predicting the overall shape of breakthrough curves (BTCs) for Ca and Mg ions, which were obtained from miscible displacement columns having different size soil aggregates. Other models that consider several mechanisms including ion exchange, complexation, dissolution/precipitation, and competitive adsorption include FIESTA (Jennings et al., 1982), CHEMTRAN (Miller and Benson, 1983) and TRANQL (Cederberg et al., 1985), among others. A major feature of these multicomponent approaches is the assumption of local equilibrium of the governing reactions. Due to their complexity, several of these models have not been fully validated, however. Miller and Benson (1983) applied CHEM-TRAN with success to the Ca and Mg data of Valocchi et al. (1981). Kirkner et al. (1985) utilized the FIESTA model to describe Ni and Cd breakthrough results on a sandy soil. Model predictions provided higher retardation of Cd and lower retardation for Ni. However, improved predictions were obtained when a kinetic approach was used with approximate parameters obtained from batch experiments.

GENERAL ASSUMPTIONS

Two reaction mechanisms of heavy metals are assumed to be the dominant retention processes in the soil system: (1) cations are adsorbed as readily exchangeable ions, and (2) cations are sorbed with high affinity to specific sites on the soil matrix. According to Sposito (1989), reactions of the first type are generally referred to as *ion exchange*. The second type of reactions includes processes that cause strong cation retention in the soil system. Such reactions may include formation of inner-sphere complexes, surface precipitation, and possibly, the penetration of heavy metal cations into the lattices of soil minerals. For convenience, we refer to the second type of reactions as *specific sorption* in a similar fashion to that used by Tiller et al. (1984).

Ion exchange is considered an instantaneous process representing (nonspecific) sorption mechanisms and as a fully reversible reaction between heavy metal ions in the soil solution and those retained on charged surfaces of the soil matrix. In contrast, specific sorption is a second-order kinetic process with high affinity for the solid phase. Because of such high affinity, retention via specific sorption is assumed irreversible or only weakly reversible. For a multi-ion system, incorporation of the two retention mechanisms, ion exchange and specific sorption, into the convective–dispersion transport equation (for a heavy metal ion i) yields:

$$\Theta \frac{\partial C_i}{\partial t} + \rho \left[\frac{\partial S_i}{\partial t} + \frac{\partial \Psi_i}{\partial t} \right] = \Theta D \frac{\partial^2 C_i}{\partial z^2} - q \frac{\partial C_i}{\partial z} \qquad (7\text{-}1)$$

where S_i and Ψ_i are the amounts retained on the exchanger surfaces and that specifically sorbed expressed as $mmol_c$ kg^{-1} soil, respectively, and C_i is the concentration in solution ($mmol_c$ L^{-1}). In addition, D is the hydrodynamic dispersion coefficient (cm^2 day^{-1}), q is Darcy's water flux density (cm day^{-1}), ρ is the soil bulk density (g cm^{-3}), and Θ is water content (cm^3 cm^{-3}), z is soil depth (cm), and t is time (day).

EQUILIBRIUM ION EXCHANGE

In a standard mass action formulation, the exchange reaction for two competing ions i and j, having valencies v_i and v_j respectively, may be written as (Sposito, 1981):

$$^TK_{ij} = \frac{(a_i^*/a_i)^{v_j}}{(a_j^*/a_j)^{v_i}} \quad (7\text{-}2)$$

where $^TK_{ij}$ denotes the thermodynamic equilibrium constant and a and a^* (omitting the subscripts) are the ion activity in soil solution and on the exchanger surfaces, respectively. Based on eq. 7-2, one can denote the parameter $^vK_{ij}$ as:

$$^vK_{ij} = \frac{^TK_{ij}}{\left[\dfrac{(\zeta_i)^{v_j}}{(\zeta_j)^{v_i}}\right]} \quad (7\text{-}3)$$

where ^vK is the Vanselow selectivity coefficient and ζ the activity coefficient on the soil surface. It is recognized that in soils, ion exchange involves a wide range of thermodynamically different sites. As a result, a common practice is to ignore the activity coefficients of the adsorbed-phase (ζ) in general. In addition, the much simpler Gaines and Thomas (1953) selectivity coefficient $^GK_{ij}$ may be used, where

$$^GK_{ij} = \frac{(\gamma_j)^{v_i}}{(\gamma_i)^{v_j}}\left(\frac{s_i}{C_i}\right)^{v_j}\bigg/\left(\frac{s_j}{C_j}\right)^{v_i} \quad (7\text{-}4)$$

This formulation is more conveniently incorporated into the dispersion–convection transport eq. 7-1. In eq. 7-4, γ_i and γ_j are dimensionless solution-phase activity coefficients (Stumm and Morgan, 1981) where

$$a_i = \gamma_i C_i \quad (7\text{-}5)$$

In addition, the terms s_i and s_j are dimensionless, representing the solid-phase concentrations expressed in terms of equivalent fraction (Rubin and James, 1973):

$$s_i = \frac{S_i}{\Omega} \tag{7-6}$$

and Ω is the cation exchange (or adsorption) capacity of the soil ($mmol_c$ kg^{-1} soil) and S_j is the concentration of adsorbed phase ($mmol_c$ kg^{-1} soil) as defined earlier in eq. 7-1. In the subsequent analysis, we consider Ω an intrinsic property representing the magnitude of negative charge on soil matrix surfaces, where

$$\Omega = \sum_i S_i \tag{7-7}$$

Although Ω is often assumed as invariant, it is recognized that Ω has been observed to be dependent on soil pH and the counter ions present in the soil (Sposito, 1989). Moreover, there are several other ways to express the adsorbed-phase concentration on a fractional basis, including that of a molar rather than an equivalent fraction (see Sposito and Mattigod, 1977). As pointed out by Rubin and James (1973) and Valocchi et al. (1981), the equivalent fraction representation is a convenient way of isotherm formulation for a generalized system of N ionic species.

BINARY HOMOVALENT SYSTEMS

For the case of a binary homovalent ions, i.e., $v_i = v_j = v$, and assuming similar ion activities in the solution phase ($\gamma_i = \gamma_j = 1$), eq. 7-4 can be rewritten as:

$$K_{ij} = K_{12} = \left[\frac{s_i}{C_i}\right] \bigg/ \left[\frac{s_j}{C_j}\right] \tag{7-8}$$

where K_{ij} represents the affinity of ion i over j (Rubin and James, 1973) or a separation factor for the affinity of ions on exchange sites (Helfferich, 1962). Valocchi (1981) stated that incorporation of γ is only necessary when the ionic strengths of corresponding experimental isotherms differ from those used in transport experiments. Therefore, under conditions of similar ionic strength, we can assume that eq. 7-8 is valid.

Rearrangement of eq. 7-8 yields the following isotherm relation for equivalent fraction of ion 1 as a function of c_1 as:

$$s_1 = \frac{K_{12}c_1}{1 + (K_{12} - 1)c_1} \tag{7-9}$$

and c_1 relative concentration (dimensionless) and C_T ($mmol_c$ L^{-1}) represents the total solution concentration:

$$c_i = \frac{C_i}{C_T} \quad and \quad C_T = \sum_i C_i \qquad (7\text{-}10)$$

The respective isotherm equation for ion 2 (i.e., s_2 vs. c_2) can be easily deduced. This dimensionless isotherm relation is represented in Figure 7-1 for different values of K_{12}. For $K_{12} = 1$, a linear isotherm relation is produced, represented by the solid line in Figure 7-1. This clearly illustrates a 1:1 relationship between relative concentration in solution and that on the adsorbed phase (i.e., $s_1 = c_1$). This also implies that the two ions 1 and 2 each have equal affinity for the exchange sites. In contrast, for $K_{12} \neq 1$, we have nonlinear sorption isotherms. Specifically, for $K_{12} > 1$, sorption of ion 1 is preferred and the isotherms are convex. For $K_{12} < 1$, sorption affinity is apposite and the isotherms are concave. Examples of homovalent ion exchange isotherms are illustrated in Figure 7-2 for Ca-Mg in soils and clay mineral systems (Gaston and Selim, 1990a,b) and in Figure 7-3 for Cd-Ca in two soils (Selim et al., 1992).

MULTIPLE ION SYSTEMS

For the general case of a multiple ion system, the adsorption of any one ion i is influenced by the solution and adsorbed-phase concentrations of the other ion species present. Therefore, adsorbed concentration S_i cannot, in general, be expressed solely as a function of its solution concentration. For the case of multiple ion exchange, Rubin and James (1973) derived a set of recursion formulas for the time derivatives of the amount adsorbed, expressed in equivalent fractions (s), in terms of the solution concentration (C) of all ionic species present in the

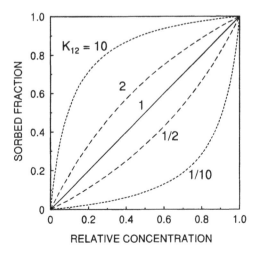

Figure 7-1 Exchange isotherms as affected by different values of selectivity coefficients (K_{12}).

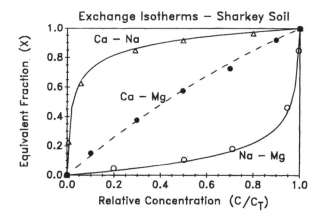

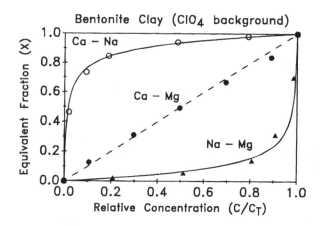

Figure 7-2 Exchange isotherms for Mg–Na, Ca–Mg, and Na–Ca on Sharkey soil (top) and bentonite-sand mixture (bottom). Smooth curves are predictions obtained using the constant exchange selectivity model.

soil solution. In adaptation, the multicomponent exchange formulation, when N ionic species are considered (Gaston and Selim, 1990a), are

$$\partial s_i/\partial t = (1/g_i)[h_{ii}(\partial C_i/\partial t) - \sum_{\substack{j=1 \\ j \neq i}}^{N} h_{ij}(\partial C_j/\partial t)] \tag{7-11}$$

where

$$g_i = 1 - \sum_{\substack{j=1 \\ j \neq i}}^{N} (\partial f_{ij}/\partial s_i)/(\partial f_{ij}/\partial s_j) \tag{7-12}$$

$$h_{ii} = \sum_{\substack{j=1 \\ j \neq i}}^{N} (\partial f_{ij}/\partial C_i)/(\partial f_{ij}/\partial s_j) \tag{7-13}$$

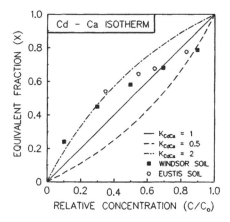

Figure 7-3   Cadmium–calcium exchange isotherm for Windsor and Eustis soils. Solid and dashed curves are simulations using different selectivities (K_{CdCa}).

$$h_{ij} = -(\partial f_{ij}/\partial C_j)/(\partial f_{ij}/\partial s_j) \tag{7-14}$$

$$f_{ij} = K_{ij}C^{vj}s_j^{vi} - C^{vi}s_i^{vj} \tag{7-15}$$

$$K_{ij} = (C_j^{vi}s_i^{vi})/(C_i^{vi}s_j^{vi}) \tag{7-16}$$

Inherent in this approach is that the generic selectivity coefficients (K_{ij}) for all cation pairs i and j, defined by eq. 7-16, is constant over all ranges of exchanger phase compositions. It is further assumed that selectivity coefficients determined from binary exchange data are applicable in higher order exchange systems. Implicit in the choice of eq. 7-16 for the selectivity coefficient are the assumptions that the equivalent fraction (s_i) adequately models surface phase activity and that the solution phase activity coefficients vary little over the range of ionic strength that occurs. An additional assumption is that the cation exchange capacity (Ω) is constant. The above formulations are applicable for systems of multiple ionic species having different valency and under variable total concentrations (C_T) of the incoming soil solution. Applications of this approach for ternary systems (Na, Ca, and Mg) have been successfully carried out by Mansell et al. (1988) and Gaston and Selim (1990a,b).

VARIABLE SELECTIVITIES

Equilibrium ion exchange as described above has been employed to describe sorption of heavy metals in soils by several investigators (Abd-Elfattah and Wada, 1981; Harmsen, 1977; Bittel and Miller, 1974; Selim et al., 1992; Hinz and Selim, 1994). In general, the affinity of heavy metals increases with decreasing heavy metal fraction on exchanger surfaces. Using an empirical selectivity coefficient, it was shown that Zn affinity increased up to two orders of magnitude for low Zn surface coverage in a Ca background solution (Abd-Elfattah and Wada,

1981). Recently, Mansell et al. (1988) relaxed the assumption of constant K_{ij} values and allowed the selectivity coefficients to vary with the adsorbed amount on the exchange (s_i). They incorporated variable selectivity into the recursion formulas of Rubin and James (1973) (eqs. 7-11 to 7-16) and successfully described the transport of Ca, Mg, and Na in a Yolo soil.

The Rothmund–Kornfeld binary exchange is another approach that incorporates variable selectivity based on the amount of adsorbed (s_i) or exchanger composition. The approach is empirical and provides a simple equation that incorporates the characteristic shape of binary exchange isotherms as a function of equivalent fraction of the amount sorbed (s_i) as well as the total solution concentration in solution (C_T). Harmsen (1977) and Bond and Phillips (1990) expressed the Rothmund–Kornfeld as:

$$\frac{(s_i)^{vj}}{(s_j)^{vi}} = {}^R K_{ij} \left[\frac{(c_i)^{vj}}{(c_j)^{vi}} \right]^n \tag{7-17}$$

where n is a dimensionless empirical parameter associated with the ion pair i-j and ${}^R K_{ij}$ is the Rothmund–Kornfeld selectivity coefficient. The above equation is best known as a simple form of the Freundlich equation that applies to ion exchange processes. As pointed out by Harmsen (1977), the Freundlich equation may be considered as an approximation of the Rothmund–Kornfeld equation valid for $s_i \ll s_j$ and $c_i \ll c_j$, where:

$$s_i = {}^R K_{ij}(c_i)^n \tag{7-18}$$

The ion exchange isotherms in Figure 7-4 show the relative amount of Zn and Cd adsorbed as a function of relative solution concentration, along with best-fit isotherms based on the Rothmund–Kornfeld equation for two acidic soils (Hinz and Selim, 1994). The diagonal line represents a non-preference isotherm (${}^R K_{ij} = 1$, $n = 1$) where competing ions (Ca-Zn or Ca-Cd) have equal affinity for exchange sites. The sigmoidal shapes of the isotherms reveal that Zn and Cd sorption exhibit high affinity at low concentrations, whereas Ca exhibits high affinity at high heavy metal concentrations. This behavior is well described by the Rothmund–Kornfeld isotherm with $n < 1$. Isotherms in Figure 7-4 are consistent with those of Bittel and Miller (1974), Harmsen (1977), and Abd-Elfattah and Wada (1981). However, Abd-Elfattah and Wada obtained much larger selectivities for low heavy metal concentrations on exchange surfaces compared to the results shown.

KINETIC ION EXCHANGE

Several studies indicated that ion exchange is a kinetic process in which equilibrium was not reached instantaneously. Sparks (1989) compiled an extensive list of cations (and anions) that exhibited kinetic ion exchange behavior in soils,

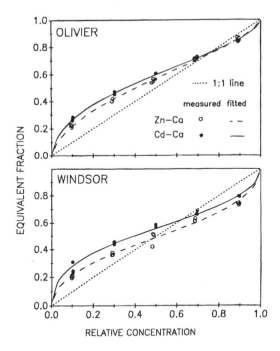

Figure 7-4 Ion exchange isotherms of Cd–Ca and Zn–Ca for Olivier and Windsor soils (relative concentration (C/C_T) versus the sorbed fraction (s/Ω). Solid and dashed curves are fitted using the Rothmund–Kornfeld equation.

e.g., Al, NH_4, K, and several heavy metal cations. According to Ogwada and Sparks (1986), observed kinetic ion exchange behavior was probably due to mass transfer (or diffusion) and chemical kinetic processes. It was postulated that in 2:1 type minerals, intra-particle diffusion is a rate-controlling mechanism governing the kinetics of adsorption of cations. Jardine and Sparks (1984) showed that the rate of K sorption–desorption (in a K-Ca system) was rapid for kaolinite and montmorillonite. However, the rate of K exchange was slow for vermiculite. Therefore, we extended the formulation described above to account for the kinetics of ion exchange. The proposed approach was analogous to mass transfer or diffusion between the solid and solution phase such that, for ion species i:

$$\frac{\partial \Gamma_i}{\partial t} = \alpha(\Gamma_i^* - \Gamma_i) \qquad (7\text{-}19)$$

where at any time t, the symbol Γ_i denotes the amount sorbed, Γ_i^* is the amount sorbed at equilibrium, and α is an apparent rate coefficient (day^{-1}) for the kinetic-type sites. In eq. 7-19, the amount sorbed at equilibrium Γ_i^* was calculated using the respective isotherm relations similar to eq. 7-9. Expressions similar to eq. 7-19 have been used to describe mass transfer between mobile and immobile water as well as chemical kinetics (Parker and Jardine, 1986). From eq. 7-19, it is obvious that for large α, Γ_i approaches Γ_i^* in a relatively short time and equilibrium is rapidly achieved. In contrast, for small α, kinetic behavior should be dominant for extended periods of time. Illustrative examples are presented in a later section.

SPECIFIC SORPTION

Several investigations discussed the role of specific sorption on the kinetic behavior of heavy metal ion retention in soils. Sorption/desorption studies showed that highly specific sorption mechanisms are responsible for metal ion retention for low concentrations. The general view was that metal ions have a high affinity for sorption sites of oxide minerals surfaces in soils. In addition, these specific sites react slowly with heavy metals and are weakly reversible. Selim et al. (1992) considered the specific sorption process ($\partial \Psi_i / \partial t$) of eq. 7-1 as a kinetic reaction where the rate of sorption is governed by a second-order mechanism. We further assumed that specific sorption occurs between metal ions present in the soil solution and that on specific sites such that:

$$\rho \frac{\partial \Psi_i}{\partial t} = k_f \Theta \phi_i C_i - k_b \rho \Psi_i = k_f \Theta (S_T - \Psi_i) C_i - k_b \rho \Psi_i \qquad (7\text{-}20)$$

where k_f and k_b are the forward and backward rate coefficients (day^{-1}), ϕ is the amount of available or vacant specific sites, and S_T is the total amount of specific sorption sites (mmol$_c$ kg^{-1}). Available or vacant specific sites are not strictly vacant. They are assumed to be occupied by hydrogen, hydroxyl, or other specifically sorbed species. As $t \rightarrow \infty$ when equilibrium is achieved, the second-order eq. 7-2 obeys the widely recognized Langmuir sorption isotherm equation (omitting the subscript i):

$$\frac{\Psi}{S_T} = \frac{\omega C}{1 + \omega C} \qquad (7\text{-}21)$$

where ω ($= \Theta k_f / \rho k_b$) is the (equilibrium) Langmuir coefficient (Sposito, 1989).

The role of specific sorption and its influence on metal ion behavior in soils has been recognized by several scientists. Sorption/desorption studies showed that specific sorption mechanisms are responsible for metal ion retention for low concentrations (Garcia-Miragaya and Page, 1976; Tiller et al., 1979, 1984). In the absence of competing metal ions for specific sites (e.g., Ni, Co, Cu, etc.), as is the case in this study, it is reasonable to consider specific sorption as an irreversible process. Therefore, the above second-order reaction was modified to describe irreversible or weakly reversible retention by setting the backward rate coefficient k_b to zero:

$$\rho \frac{\partial \Psi_i}{\partial t} = k_f \Theta(S_T - \Psi_i)C_i \tag{7-22}$$

As a result, only two parameters, S_T and k_f, are required to account for irreversible retention. For several metal ions including Cd, Ni, Co, and Zn, specific sorption has been shown to be dependent on time of reaction. Therefore, the use of a kinetic rather than an equilibrium sorption mechanism is recommended. In eq. 7-22, the total amount of specific sites S_T was found to be highly dependent on the type of surface sites and pH (Abd-Elfattah and Wada, 1981; Tiller et al., 1984). Moreover, a major advantage of the formulation of irreversible reaction eq. 7-22 is that a sorption maximum is achieved when all unfilled sites become occupied.

SIMULATIONS

Figures 7-5 to 7-10 represent several simulations using the competitive ion exchange approach. The simulations illustrate the fate of a heavy metals in soil when ion exchange and specific sorption mechanisms are the dominant retention reactions. In Figure 7-5, the effect of K_{12} on ion transport is represented by the breakthrough curves (BTCs) shown. These are represented by the relative concentration (for ion 1) when plotted versus the relative pore volume (V/V_o). The BTC illustrated by the solid curve is for a homovalent ion system having equal affinity to the exchange sites ($K_{12} = 1$). The dotted curve is for $K_{12} = 1/10$, which reflects strong affinity of ion 2 over ion 1 for exchange sites. For this case, there was little retention of ion 1, as indicated by the early arrival of the BTC. In fact, the BTC resembles that of a nonreactive solute. In contrast, for the case where $K_{12} = 10$ (dashed BTC), a high degree of retention was

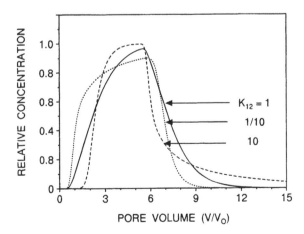

Figure 7-5 Simulated BTCs for several K_{12} values under constant total solution concentration (C_T).

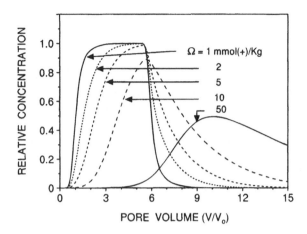

Figure 7-6 Simulated BTCs for several Ω values under constant total solution concentration (C_T).

observed. This is illustrated by the retardation of the BTC and extensive spreading of the effluent.

Model parameters used to obtain the simulations shown in Figure 7-5 include: $\Theta = 0.4$ cm^3 cm^{-3}, $\rho = 1.2$ g cm^{-3}, $q = 1$ cm day^{-1}, $D = 1$ cm^2 day^{-1} $L = 10$ cm, Peclet number ($P = qL/\Theta D$) = 25, $\alpha = 0.5$ day^{-1}, $k_f = 0$, and $\Omega = 10$ mmol$_c$ kg^{-1}. Moreover, the total concentration (C_T) was assumed constant and a pulse of solution (5 pore volumes) containing only ion 1 was introduced into a column replacing a solution containing only ion 2. These model parameters were also used to obtain the simulations shown in Figure 7-6. Here, the effects of the exchange capacity or total amount of sites (Ω) on the shape of the BTC are clearly illustrated. A value of 2 for K_{12} was used here and for subsequent simulations. Model sensitivity to the kinetic rate coefficient (α) is illustrated by the BTCs shown in Figure 7-7, where a range of α values was chosen. As α

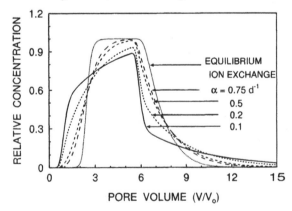

Figure 7-7 Simulated BTCs for different values of α under constant total solution concentration (C_T).

increases, i.e., when ion exchange becomes more rapid or is considered instantaneous in nature ($\alpha \to \infty$), the BTCs show a sharp rise of the influent side and little tailing of the effluent (desorption) side. Maximum values for peak concentrations were achieved when conditions of local equilibrium prevailed. As ion exchange becomes more kinetic (i.e., as α decreases), the BTCs become less retarded, as indicated by the early arrival of ion 1 in the effluent. This is a direct result of the influence of α on the residence time of the ions in the soil column. The simulations in Figure 7-7 also reveal that maximum (peak) concentrations of the BTCs increased as α increased. Moreover, spreading of the effluent, as manifested by tailing of the desorption side of the BTCs, was observed for decreasing values of the kinetic parameter α.

The simulations shown in Figures 7-8 to 7-10 represent miscible displacement cases where the total solution concentration (C_T) of the solute pulse was allowed to vary with time. For C_2 at any depth and time, the results were simply obtained from the appropriate solutions for C_T and C_1. Here, a solute pulse devoid of C_2 with $C_T = C_1 = 1$ mmol$_c$ L^{-1} was applied to a 10-cm soil column having $C_T = C_2 = 20$ mmol$_c$ L^{-1}. Following the C_1 pulse, the applied solute was $C_T = C_2 = 20$ mmol$_c$ L^{-1}. All other parameters, such as Ω, q, D, etc., were the same as given above. Simulation results of Figure 7-8 show that the influence of α on the shape and location of the BTCs for ion 1 are similar to those of Figure 7-7, where C_T was considered time invariant. The major exception is that concentration maxima were as much as threefold that of the C_1 concentration in the applied pulse (1 mmol$_c$ L^{-1}). This phenomena is referred to as the *chromatographic effect* or commonly called a *snow plow* (Starr and Parlange, 1979) and is due primarily to large differences between input (pulse) and initial concentration. The snow plow effect is also dependent on the relative magnitudes of Ω and C_T becoming more pronounced as the ratio of C_T to Ω increases. Figure 7-9 shows the BTCs for C_T and C_2 for the corresponding BTCs shown in Figure 7-8. For all simulations, the BTC for C_T remained the same, with the effect of

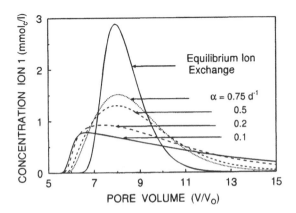

Figure 7-8 Simulated BTCs for different values of α under conditions of variable total solution concentration (C_T).

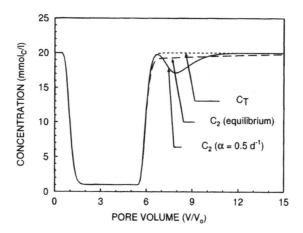

Figure 7-9 Simulated BTCs for C_T and C_2 for equilibrium and kinetic ion exchange (α = 0.5 day^{-1}).

α manifested by the shape of C_2 (and C_1) BTC only. In order to illustrate the effect of the specific sorption process on the shape of the BTCs shown in Figure 7-10, we selected the cases where C_T was allowed to vary under conditions of local equilibrium for the ion exchange process with all remaining parameters the same as listed above. It is clear that the effect of irreversible retention is that of an overall decrease of solute concentrations. As k_f increased, the peak maxima decreased, with the amount irreversibly sorbed due to the specific sorption mechanism, represented by the area between any BTC curve and that where specific sorption was not incorporated ($k_f = 0$).

APPLICATIONS

Several examples are given here to illustrate the capability of the competitive model in describing heavy metals transport in soils. Figure 7-11 is an example

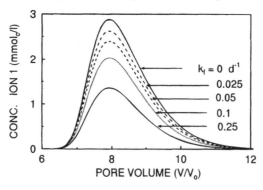

Figure 7-10 Simulated BTCs for different values of k_f under conditions of variable total solution concentration (C_T).

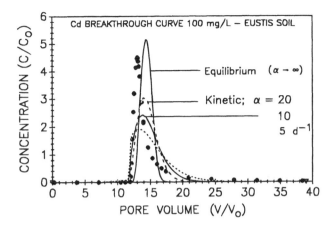

Figure 7-11   Measured (closed circles) and predicted breakthrough curves in Windsor soil column. Curves are predictions using equilibrium and kinetic ion exchange with different α values.

of Cd miscible displacement results for Eustis fine sandy soil from Selim et al. (1992). The sequence of input solutions was 10 mmol$_c$ L^{-1} of Ca(NO$_3$)$_2$, followed by 1.786 mmol$_c$ L^{-1} Cd(NO$_3$)$_2$, then 10 mmol$_c$ L^{-1} of Ca(NO$_3$)$_2$ subsequently added. As a result of changes in the total concentration (C_T), a pronounced snow plow effect was observed, as shown in Figure 7-11. Here, the measured peak concentration was about fivefold the input Cd pulse. Moreover, use of the ion exchange approach with the assumption of local equilibrium adequately predicted the shape and location of the BTC, although the BTC was somewhat more retarded than observed. The selectivity coefficient (K_{12}) used in the simulations shown in Figure 7-11 was equal to 1, indicating equal affinity of Cd and Ca to the soil surface. It is clear that the use of the kinetic ion exchange improved overall prediction of the BTC even though this resulted in the lowering of peak concentrations. Therefore, the competitive ion exchange approach was capable of predicting the snow plow elution of the Cd pulse from the soil column.

To further test the capability of the competitive model, two data sets from multiple pulse applications are illustrated. Figures 7-12 and 7-13 are for Windsor soil where Cd pulse applications were 10 and 100 mg L^{-1}, respectively (Selim et al., 1992). For all multiple pulses, the ion exchange approach well predicted the position of the BTC peaks. In fact, the assumption of equilibrium ion exchange adequately predicted the observed snow plow effect for the two Windsor data sets. Good predictions were also obtained for peak maxima for the 100 mg L^{-1} data set (Figure 7-13). Calculated BTCs shown in these figures were obtained from input parameters that were independently determined. Specifically, curve-fitting of the data was not implemented to obtain the predictions shown. The shape of the Cd peaks of Figures 7-12 and 7-13 are due in part to the concentration and width or the number of pore volumes of each input pulse (Selim et al., 1992). Lower peak concentrations for 10 mg L^{-1} in comparison to 100 mg L^{-1} are perhaps related to the equivalent fraction of Cd on exchange surfaces for Windsor

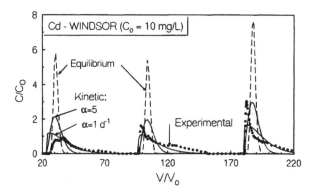

Figure 7-12 Measured (closed circles) and predicted breakthrough curves in Windsor soil
column for three Cd pulses of C_o = 10 mg L^{-1}. Curves are predictions using
equilibrium and kinetic ion exchange with different α values.

soil. In Figure 7-12, the maximum C/C_o increased from 0.9 to 1.6 and reached
α C/C_o of 3.0 for the third Cd pulse. The first pulse (22.6 pore volumes) represents
about 4% of the total cation exchange in the soil column and the second and
third pulses (20.8 and 30.3 pore volumes) represent 4 and 6%, respectively. Selim
et al. (1992) postulated that applied Cd may occupy specific sorption sites on
matrix surfaces. Therefore, irreversible Cd sorption could partly explain the fact
that only 80% of applied Cd was recovered in the effluent. For the Windsor
BTCs shown in Figure 7-13 (C_o = 100 mg L^{-1}), the first pulse alone was
equivalent to 30% of the cation exchange capacity of the soil column and perhaps
resulted in higher peak maximas due to Ca-Ca ion exchange.

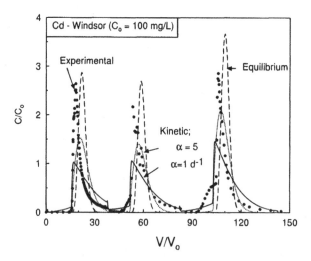

Figure 7-13 Measured (closed circles) and predicted breakthrough curves in Windsor soil
column for three Cd pulses of C_o = 100 mg L^{-1}. Curves are predictions
using equilibrium and kinetic ion exchange with different α values.

Examples of transport behavior of Zn when variable ionic strength (or C_T) conditions prevailed in the soil columns are presented in Figures 7-14 and 7-15 for two flow velocities (Hinz and Selim, 1994). Since the total concentration of the Zn and Cd input pulse solutions were much lower than that of the displacing Ca solution, chromatographic peaks were observed after 13 pore volumes. The BTCs were somewhat similar in shape except for earlier Zn arrival at high flow velocity. Early appearance of Zn was well described by the predicted BTC (dashed curves) where equal Ca-Zn exchange affinity was assumed. In fact, the chromatographic effect for Ca and Zn was adequately described by the equal affinity BTCs, for both flow velocities. However, the tailing was not well predicted. Contrary to our expectations, the Rothmund-Kornfeld equation predictions (solid curves), for both data sets, were disappointing. The extent of Zn retardation was overestimated by 2 to 3 pore volumes (right shift). The opposite trend was observed for Ca, where the BTCs were shifted to the left of experimental results. These predictions are indicative of strong Zn affinity (compared to Ca) at low Zn concentrations based on parameter estimates of ion-exchange isotherms using the Rothmund–Kornfeld approach.

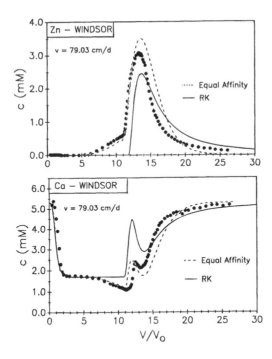

Figure 7-14 Zn and Ca breakthrough curves in Windsor soil column at variable ionic strength. Predictions were based on equal affinity ($K_{12} = 1$) and the Rothmund–Kornfeld (RK) equation.

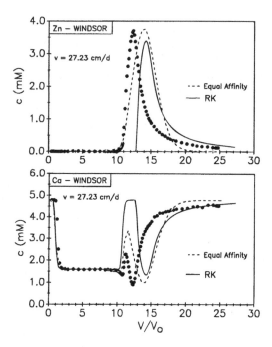

Figure 7-15 Zn and Ca breakthrough curves in a Windsor soil column at variable ionic strength. Predictions were based on equal affinity ($K_{12} = 1$) and the Rothmund–Kornfeld (RK) approach.

8 RELATIONSHIP BETWEEN MODELS AND SOIL CHEMICAL REACTIONS

The model applications presented in the previous chapters centered on describing the kinetics and transport of heavy metals in soils without considering the specific reactions involved or the soil properties that influence retention. This is because the models are mass-action formulations that do not depend on a particular reaction mechanism. They are applicable to a wide array of reactive processes involving metals in soils. In this chapter, we relate the models to specific chemical reactions and other processes between metals and soil components. We also discuss guidelines for selecting which model to use for a particular application and alternate approaches for estimating some of the key model parameters.

It is not possible to deduce reaction mechanisms that occur when soils react with solutes based solely on model predictions of the loss of solutes from aqueous solutions. Alternative processes may give rise to the same experimental observations. This point was discussed by Skopp (1986) in his review of time-dependent processes in soils. Sposito (1986) pointed out that the complexity of the processes that occur when ions are retained by soils defies unambiguous interpretations by simple models. The physical and chemical processes that comprise reaction mechanisms occur at the molecular level, whereas experimental observations on soil systems are usually at the macroscopic level. Furthermore, reaction mechanisms can never really be proved, only disproved if suitable experiments can be designed to rule out unambiguously a particular mechanism. Despite these difficulties, it is possible to propose mechanisms that are consistent with the models and experimental observations.

ION EXCHANGE AND SURFACE COMPLEXATION KINETICS

Numerous experimental studies reported in the literature help to relate the models to the actual reactions of heavy metals in soils. It is well established that cationic metal species are sorbed by permanent-charge soil components such as the smectite group of clay minerals as well as by variable-charge components such as organic matter and metal oxides (Kinniburgh and Jackson, 1981; Sposito, 1984). Harter (1984) proposed that the most rapid reversible retention reaction of metal ions in soils was nonspecific ion exchange, whereas the slower retention reaction was specific sorption of metal ions by soil surfaces. These reactions may also be interpreted in terms of formation of outer- and inner-sphere complexes with soil surfaces (Sposito, 1984). This interpretation is entirely compatible with the two-site models presented in this book.

Harter's experimental data was taken at very short time intervals using a batch reactor (Zasoski and Burau, 1978). The method used in our studies is only suitable for highly time-dependent or slow reactions. However, the applicability of the model is independent of the time scale of the experiments, since there is probably a continuum of reactions covering a wide time period. The model curve must still pass through the data points regardless of when measured, although the fastest reactions may be complete before the first data point is measured.

Sorption of cationic species by permanent-charge clays is best described by a cation exchange model and this is considered in Chapter 7. Sorption of cationic species by variable-charge surfaces is best described by a surface complexation approach that explicitly accounts for surface charge, pH, and ionic strength effects (Dzombak and Morel, 1990). Surface complexation of a metal cation by a hydroxyl group on the surface of a metal oxide involves release of surface protons:

$$SOH(s) + M^{2+}(aq) = SOM^+(s) + H^+(aq) \qquad (8\text{-}1)$$

where $SOH(s)$ is the metal oxide surface hydroxyl group, $M^{2+}(aq)$ is the metal cation in aqueous solution, and $SOM^+(aq)$ is the metal cation–metal oxide surface complex. The surface complex formed may be inner-sphere or outer-sphere, depending on the properties of the metal cation.

The rate of loss of metal cation from solution is equal to the rate of sorption on the surface and is given by:

$$\frac{d[M^{2+}]}{dt} = -k_1[M^{2+}][SOH] + k_2[SOM^+] \qquad (8\text{-}2)$$

where the brackets refer to concentrations. Formation of the surface complex is a second-order reaction. If the concentration of surface sites is substantially greater than the concentration of cations available for binding, then eq. 8-2 reduces to a pseudo first-order reaction. If the rate of sorption depends on some fractional power of the cation concentration in aqueous solution, then eq. 8-2 reduces to the nonlinear model presented in Chapter 3.

Oxyanion metals such as Cr(VI) species are also sorbed by variable-charge soil components. Buchter et al. (1989) found that Cr(VI) retention in a group of soils of differing properties was primarily influenced by pH and metal oxide content. Chromium(VI) retention decreased with increasing pH and increased with metal oxide content. The Freundlich parameter n was found to vary linearly with pH. The principal soil minerals that sorb the chromate oxyanion include kaolinite (Zachara et al., 1988) and iron oxides (hydrous Fe oxide, goethite,

Al-substituted goethite) (Zachara et al., 1987; Ainsworth et al., 1989). Burden (1989) found that hydrous Al oxide and humic acid sorbed relatively little amounts of Cr(VI) compared to goethite. Of the three soils presented in Chapters 3 and 5 as examples of the kinetics of Cr(VI) retention, Cecil soil had the highest iron oxide content and also retained the most Cr(VI) (Amacher and Selim, 1994).

Oxyanion (e.g., chromate) retention by variable surface charge minerals such as iron oxides is believed to occur by a ligand exchange mechanism (Sposito, 1984):

$$SOH(s) + H^{+(aq)} = SOH^+(s) \tag{8-3}$$

$$SOH_2^+(s) + L^{n-}(aq) = SOH_2^+ L^{n-}(s) \tag{8-4}$$

$$SOH_2^+ L^{n-}(s) = SL^{1-n}(s) + H_2O(l) \tag{8-5}$$

where SOH is the metal oxide surface hydroxyl group, SOH_2^+ is the protonated surface hydroxyl group, L is the oxyanion ligand of valance n, $SOH_2^+ L^{n-}$ is the outer-sphere metal–ligand surface complex, and SL^{1-n} is the inner-sphere metal–ligand surface complex. The first step in the reaction sequence is protonation of the surface hydroxyl group. With some weakly bound oxyanions (e.g., nitrate), the reaction may stop at the outer-sphere complex. With other more strongly bound oxyanions (e.g., phosphate), the outer-sphere complex may simply be a reactive intermediate and the steady-state approximation may be used to describe the kinetics of its formation and conversion. The rate of loss of L^{n-} from aqueous solution is

$$\frac{d[L^{n-}]}{dt} = -k_1[SOH_2^+][L^{n-}] + k_2[SOH_2^+ L^{n-}] \tag{8-6}$$

Note that eq. 8-6 is also second order and is thus fully consistent with the second-order model presented above. If more than one type of retention site exists, then eq. 8-6 is written for each type of site:

$$\frac{d[L^{n-}]}{dt} = -k_1[SOH_2^+]_1[L^{n-}] + k_2[SOH_2^+ L^{n-}]_1 - k_3[SOH_2^+]_2[L^{n-}]$$

$$+ k_4[SOH_2^+ L^{n-}]_2 \tag{8-7}$$

where $[SOH_2^+]_1$ and $[SOH_2^+]_2$ refer to type 1 and type 2 sites, respectively. Thus, the reversible reactions between solute in aqueous solution and sites ϕ_1 and ϕ_2 in the soil phase (eq. 5-5 and 5-6) are consistent with the surface complexation approach.

The rate coefficients given in eq. 8-2, 8-6, and 8-7 are conditional rate coefficients that depend on the surface charge of the sorbing surface and the pH and ionic strength of the electrolyte solution in contact with the sorbing surface.

Conditional rate coefficients can be related to intrinsic rate coefficients that account for surface charge effects:

$$k_{int} = k_c \exp\left[\frac{\Delta ZF\Psi}{RT}\right]$$
(8-8)

where k_{int} is the intrinsic rate coefficient, k_c is the conditional rate coefficient, ΔZ is the net change in charge of the surface complex, F is the Faraday constant, Ψ is the surface potential, R is the molar gas constant, and T is absolute temperature.

SURFACE COMPLEXATION EQUILIBRIUM

At equilibrium, solute retention by variable-charge surfaces can be described by any of a number of surface complexation models that account for the effect of surface charge on solute sorption (Sposito, 1984). The more popular surface complexation models in use today include the constant capacitance model, the diffuse-layer model, which is very similar to the constant capacitance model, and the triple-layer model (Sposito, 1984; Dzombak and Morel, 1990). The primary difference among these models is in their molecular description of surface complexation reactions and the interface between the variable-charge surface and aqueous solution in contact with the surface. All of these models are more or less equally capable of describing macroscopically measured sorption data despite their different molecular descriptions of the surface-solution interface. This is because macroscopic sorption data are insensitive to the molecular structure of the interfacial region (Sposito, 1984). However, Toner and Sparks (1995) recently showed that only the triple-layer model could describe both equilibrium and kinetic data for B sorption on alumina assuming $B(OH)_3$ and $B(OH)_4^-$ were adsorbed by ligand exchange on neutral surfaces. The constant capacitance and diffuse-layer models were not successful in modeling both the equilibrium and kinetic data.

In some cases of heavy metal contamination of soils, a particular variable-charge soil component such as an iron oxide will dominate sorption of the metal of interest. In such cases, it is appropriate to model both kinetic and equilibrium sorption using a surface complexation approach, provided the necessary surface complexation constants and sorbing surface characterization data are available to insure accurate modeling. Examples of surface complexation modeling of heavy metal sorption by metal oxides in soils are scarce, although Zachara et al. (1989) used the triple-layer model to calculate sorption of Cr(VI) by subsoils. Kent et al. (1995) used the diffuse-layer model to calculate sorption of Cr(VI) by metal oxides in aquifer material. They found that the diffuse-layer model overpredicted Cr(VI) sorption by aquifer material using equilibrium constants for anion sorption by hydrous ferric oxide (ferrihydrite). The vast majority of examples in the literature of surface complexation modeling of solute sorption are with pure metal oxide suspensions. All too often, the necessary information for surface complexation modeling of solute sorption in soils is lacking and the modeler is forced to fall back on a more empirical approach such as using the Freundlich or Langmuir equations.

Aside from their inability to provide any information as to the actual processes of metal retention in soils, one of the chief criticisms of the continued use of the Freundlich or Langmuir equations is that they do not explicitly consider the effects of pH and ionic strength on metal sorption by variable-charge surfaces as do the surface complexation models. This deficiency can be overcome in part by measuring sorption isotherms over the range of pH and ionic strength values that are expected in a particular metal contamination scenario. In many situations, soil pH will be buffered to a rather limited range and the effect of variable pH on metal sorption need not be considered.

It is also possible to empirically relate Freundlich and Langmuir sorption parameters to important soil properties such as pH. For example, Buchter et al. (1989) found that the Freundlich parameter b (eq. 1-1), which we use as an estimate of the reaction order in the nonlinear multireaction model in Chapter 3, showed a strong positive correlation to soil pH for sorption of the cationic metals Co, Ni, Zn, and Cd by several soils and a strong negative correlation to soil pH for sorption of the chromate oxyanion (Figure 8-1). Burden (1989) found that the Freundlich parameter n and the two-site Langmuir parameters F and S_{max} for Cr(VI) retention by the variable surface charge mineral goethite could be empirically related to the pH of the goethite suspension. Thus, it is still valid to use empirical sorption equations such as the Freundlich and Langmuir equations in soils dominated by variable-charge minerals, although the surface complexation approach is to be preferred when the necessary constants and other needed characterization data are available. Neither the surface complexation approach or a strictly empirical sorption model provide any information as to the actual

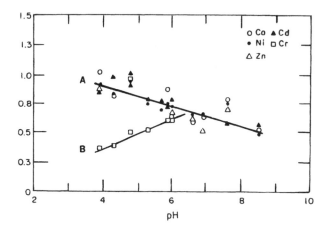

Figure 8-1   Correlation between soil pH and Freundlich parameter n. Curve A is a regression line for Co, Ni, Zn, and Cd ($b = 1.24 - 0.0831$ pH, $r = 0.83**$). Curve B is for Cr(VI) ($b = -0.0846 + 0.116$ pH, $r = 0.98**$).

mechanism of retention. Thus, empirically measuring sorption over the range of pH and ionic strength of interest is probably no worse than the inaccuracies introduced by the assumptions and estimates needed to use a surface complexation model for a soil. The choice then becomes one based on the availability of the necessary data or of convenience.

MODEL SELECTION

If an empirical sorption equation such as the Freundlich or Langmuir is used instead of a surface complexation model, it is helpful to have some guidelines as to which model to choose for a particular application. Both the Freundlich and Langmuir equations are specialized cases of the more general expression for solute retention in soil (Goldberg and Sposito, 1984):

$$S = \sum_{i=1}^{m} \frac{S_{Ti} K_i^{n_i} C^{n_i}}{1 + K_i^{n_i} C^{n_i}}$$

(8-9)

where S is the quantity of solute sorbed per unit mass of soil (mg kg^{-1}), C is the concentration of solute in aqueous solution in contact with the soil (mg L^{-1}), and m, S_{Ti}, K_i, and n_i are empirical fitting parameters. For $m = 1$ and $n_1 = 1$, eq. 8-9 reduces to the Langmuir equation (5-14). For $m = 2$ and $n_1 = n_2 = 1$, eq. 8-9 becomes the two-site Langmuir equation (5-12). For $m = 1$, $0 < n_i < 1$, and $K_1^{n_1} C^{n_1} << 1$, eq. 8-9 becomes the Freundlich equation (3-4).

The choice of which model to use is usually based on goodness-of-fit criteria. Some additional guidelines for model selection can be given, however. If the quantity of solute is small relative to the quantity of sorption sites so that the sorption rate is not limited by the availability of sorption sites, then the nonlinear model presented in Chapter 3 is adequate. A plot of C/C_o versus time for several initial solute concentrations can indicate if the quantity of sorption sites is rate limiting. If the curves overlie one another so that the sorption rate does not depend on the magnitude of C_o, then the quantity of sorption sites is probably not rate limiting. This is likely only for very small values of C_o. In such cases, the reaction order n can be estimated by fitting the Freundlich equation to the sorption data after the reaction approaches apparent equilibrium at longer times. Even if the quantity of sorption sites is rate limiting, the nonlinear model can still fit the data. However, a different set of rate coefficients is required for each C_o.

If the curves in a plot of C/C_o vs. time for a series of C_o values do not overlie one another, then the quantity of sorption sites is probably rate limiting and the second-order model is more applicable. The choice of whether to use a one- or two-site model will usually be based on whether including an additional site will improve the fit of the model to the data. Estimates of F and S_T in the second-order model can be made by fitting the two-site Langmuir equation to the sorption data after apparent equilibrium is attained. For more information on the relationships of the Freundlich and Langmuir equations to the general sorption eq. 8-9, refer to Sposito (1984).

IRREVERSIBLE MECHANISMS

Inclusion of an irreversible first-order reaction in the models presented in Chapters 3 and 5 was justified on the basis of the continuing loss of solute from solution, which followed first-order kinetics for a given C_o after about 24 h of reaction time, and because the release of solute was not fully reversible when the solution in contact with the soil was diluted (Amacher et al., 1988). Amacher et al. (1986) proposed that the irreversible or very slowly reversible retention of Cr(VI), Cd, and Hg by soils may be precipitation, internal diffusion, or in the case of Cr(VI) and Hg, a change in the chemical species (oxidation/reduction).

In the case of Cr(VI), the irreversible reaction is most likely the reduction of Cr(VI) by organic matter (Bartlett and Kimble, 1976; James and Bartlett, 1983; Amacher and Baker, 1982) and/or minerals containing Fe^{2+} (Eary and Rai, 1991). The reduction of Cr(VI) by dissolved organic carbon (fulvic acid) is first order with respect to Cr(VI) at a constant pH, and the overall reaction rate is pH dependent (Amacher and Baker, 1982). At higher Cr(VI) concentrations, concentrations of reducing agents may become limiting, which may account for the systematic decrease in k_s with C_o that was observed by Amacher et al. (1988). Recently, Wittbrodt and Palmer (1995), in a series of kinetic experiments, worked out the complete rate equation for Cr(VI) reduction by soil fulvic acid:

$$R = k_r[HCrO_4^-][SFA][Cr(VI)]_o^p[H^+]^q \qquad (8\text{-}10)$$

where R is the rate of the reduction reaction, k_r is the reduction rate coefficient and equal to kX_e^n, $[HCrO_4^-]$ is the concentration of the chromate species, [SFA] is the concentration of soil fulvic acid, X_e is the fraction of SFA reducing equivalents oxidized, $[Cr(VI)]_o$ is the initial Cr(VI) concentration, $[H^+]$ is the hydrogen ion concentration, and k, n, p, and q are constants. The values of these constants are $k = 4.27 \pm 0.68$, $n = -0.94 \pm 0.04$, $p = 0.47 \pm 0.11$, and $q = 0.45 \pm 0.03$. This complex rate equation shows that there will be a systematic variation in k_s with C_o in our model.

Mendoza and Barrow (1987) proposed that the continuing reaction between phosphate and soils is the penetration of adsorbed phosphate into the adsorbing surface. This is an internal diffusion process, which was also proposed by Aringhieri et al. (1985) to explain why a single second-order reversible reaction could not adequately describe Cu and Cd retention kinetics by a soil. Surface precipitation, internal diffusion, and other processes have also been used to explain the continuing slow reaction of solutes with mineral surfaces (Davis and Hayes, 1986). In the case of porous minerals such as ferrihydrite, diffusion inside aggregates may account for the continuing slow uptake of solute by the mineral (Fuller et al., 1993). The importance of mineral pore size and structure on the kinetics of solute sorption by mineral surfaces was recently demonstrated by Papelis et al. (1995). They showed that Cd and selenite sorption by three porous aluminas is controlled by intraparticle diffusion. Solute sorption by the smaller particles was much faster than sorption by the larger particles.

Zhang and Sparks (1989, 1990a,b) studied the kinetics of molybdate, sulfate, selenate, and selenite adsorption/desorption on goethite using a pressure-jump relaxation method with conductivity detection. Oxyanion retention by goethite, a nonporous mineral, is extremely rapid and the rate coefficients they measured were considerably higher than those calculated by the second-order, two-site model for chromate retention by soils. Because relaxation methods measure only the kinetics of the surface complexation reaction, the difference can probably be attributed to the fact that mass transfer processes (diffusion) limit reaction rates in soils. The relaxation studies provide additional evidence that rate coefficients from batch reactor experiments with porous minerals and soils are apparent rate coefficients that include some mass-transfer processes. In such cases, kinetic models and parameters that describe **physical transport-controlled** processes must be used to analyze the data (Sposito, 1994).

Soils consist of numerous porous and nonporous minerals and organic matter. Thus, a multireaction model that contains reversible reactions consistent with surface complexation type reactions and a sink term that accounts for the continuing slow sorption of solute appears justified. Barrow (1987) also interprets the initial rate of sorption of a solute by solid surfaces as a second-order kinetic process involving adsorption of the solute by a charged surface, followed by a slower internal diffusion process into the mineral structure. The exact nature of the reactions will of course depend on the solute and reacting surfaces in the soil. Actual identification of the chemical species and reaction sites is necessary to distinguish among the various possibilities, and such independent experimental evidence is for the most part lacking in the case of reactions at soil surfaces. However, recent advances in instrumental methods that allow direct observation of chemical species in mineral surfaces and therefore determination of reaction mechanisms promise to close this information gap (AACES, 1995).

REFERENCES

AACES. 1995. *The Analyst*. Issue 1. Advanced Analytical Center for Environmental Sciences. Aiken, SC.

Abd-Elfattah, A. and K. Wada. 1981. Adsorption of lead, copper, zinc, cobalt, and calcium by soils that differ in cation-exchange materials, *J. Soil Sci.* 32:271–283.

Adriano, D. C. 1986. *Trace Elements in the Terrestrial Environment*. Springer-Verlag, New York.

Aharoni, C. and D. L. Sparks. 1991. Kinetics of soil chemical reactions—A theoretical treatment. In: *Rates of Soil Chemical Processes*. D. L. Sparks and D. L. Suarez (Eds.) SSSA Spec. Publ. No. 27. Soil Sci. Soc. Am., Madison, WI. p. 1–18.

Aharoni, C. and Y. Suzin. 1982a. Application of the Elovich equation to the kinetics of occlusion. 1. Homogeneous microporosity. *J. Chem. Soc., Faraday Trans. I.* 78:2313–2320.

Aharoni, C. and Y. Suzin. 1982b. Application of the Elovich equation to the kinetics of occlusion. 2. Analysis of experimental data from the literature. *J. Chem. Soc., Faraday Trans. I.* 78:2321–2327.

Ainsworth, C. C., D. C. Girvin, J. M. Zachara, and S. C. Smith. 1989. Chromate adsorption on goethite: effects of aluminum substitution. *Soil Sci. Soc. Am. J.* 53:411–418.

Alesii, B. A., W. H. Fuller, and M. V. Boyle. 1980. Effect of leachate flow rate in metal migration through soils. *J. Environ. Qual.* 9:119–126.

Altmann, R. S. and J. O. Leckie. 1987. Metal binding in heterogeneous multicomponent systems: mathematical and experimental modeling. In: *Oceanic Processes in Marine Pollution, Vol. 2: Physicochemical Processes and Wastes in the Ocean*. T. P. O'Connor, W. V. Burt, and I. W. Duedall (Eds.) Robert E. Krieger, Malabar, FL. p.145–157.

Amacher, M. C., and D. E. Baker. 1982. *Redox reactions involving chromium, plutonium, and manganese in soils*. DOE/DP/04515-1. U.S. Dept. Energy, Las Vegas, NV and Institute for Research on Land and Water Resources, Pennsylvania State University, University Park, PA.

Amacher, M. C., J. Kotuby-Amacher, H. M. Selim, and I. K. Iskandar. 1986. Retention and release of metals by soils—evaluation of several models. *Geoderma.* 38:131–154.

Amacher, M. C. and H. M. Selim. 1994. Mathematical models to evaluate retention and transport of chromium(VI) in soil. *Ecol. Mod.* 74:205–230.

Amacher, M. C., H. M. Selim, and I. K. Iskandar. 1988. Kinetics of chromium(VI) and cadmium retention in soils; a nonlinear multireaction model. *Soil Sci. Soc. Am. J.* 52:398–408.

Amacher, M. C., H. M. Selim, and I. K. Iskandar. 1990. Kinetics of mercuric chloride retention by soils. *J. Environ. Qual.* 19:382–388.

Amoozegar-Fard, A., A. W. Warrick, and W. H. Fuller. 1983. A simplified model for solute movement through soils. *Soil Sci. Soc. Am. J.* 47:1047–1049.

Amoozegar-Fard, A., W. H. Fuller, and A. W. Warrick. 1984. An approach to predicting the movement of selected polluting metals in soils. *J. Environ. Qual.* 13:290–297.

Anderson, M. A. and A. J. Rubin. 1981. *Adsorption of Inorganics at Solid-Liquid Interfaces.* Ann Arbor Science, Ann Arbor, MI.

Andreini, M. S. and T. S. Steenhuis. 1990. Preferential paths of flow under conventional and conservation tillage. *Geoderma.* 46:85–102.

Aringhieri, R., P. Carrai, and G. Petruzzelli. 1985. Kinetics of Cu and Cd adsorption by and Italian soil. *Soil Sci.* 139:197–204.

Barber, S. A. 1984. *Soil Nutrient Bioavailability: A Mechanistic Approach.* Wiley-Interscience, New York p. 398.

Barrow, N. J. and T. C. Shaw. 1975. The slow reactions between soil and anions. II. Effect of time and temperature on the decrease in phosphate concentration in the soil solution. *Soil Sci.* 119:167–177.

Barrow, N. J. and T. C. Shaw. 1979. Effect of solution:soil ratio and vigour of shaking on the rate of phosphate adsorption by soil. *J. Soil Sci.* 30:67–76.

Barrow, N. J. 1987. *Reactions with Variable-Charge Soils.* Martinus Nijhoff, Dordrecht, The Netherlands.

Barrow, N. J. 1989. Suitability of sorption-desorption methods to simulate partitioning and movement of ions in soils. *Ecol. Stud.* 74:3–17.

Bartlett, R. J. and J. M. Kimble. 1976. Behavior of chromium in soils. II. Hexavalent forms. *J. Environ. Qual.* 5:383–386.

Bear, J. 1972. *Dynamics of Fluids in Porous Media.* Elsevier, New York.

Bernasconi, C. F. (Ed.) 1986. *Investigations of Rates and Mechanisms of Reactions.* 4th ed. Wiley, New York.

Bittel, J. E. and R. J. Miller. 1974. Lead, cadmium and calcium selectivity coefficients of a montmorillonite, illite and kaolinite. *J. Environ. Qual.* 3:250–253.

Boast, C. W. 1973. Modeling the movement of chemicals in soils by water, *Soil Sci.* 115:224–230.

Bond, W. J. and I. R. Phillips. 1990. Approximate solution for cation transport during unsteady, unsaturated soil water flow. *Water Resour. Res.* 26:2195–2205.

Boyd, G. E., A. W. Adamson, and L. S. Meyers, Jr. 1947. The exchange adsorption of ions from aqueous solutions by organic zeolites. II. Kinetics. *J. Am. Chem. Soc.* 69:2836–2848.

Brenner, H. 1962. The diffusion model of longitudinal mixing in beds of finite length: numerical values. *Chem. Eng. Sci.* 17:220–243.

Brusseau, M. L., R. E. Jessup, and P. S. C. Rao. 1989. Modeling the transport of solutes influenced by multiprocess nonequilibrium. *Water Resour. Res.* 25:1971–1988.

Buchter, B., B. Davidoff, M. C. Amacher, C. Hinz, I. K. Iskandar, and H. M. Selim. 1989. Correlation of Freundlich K_d and n retention parameters with soils and elements. *Soil Sci.* 148:370–379.

Bunzl, K. 1974. Kinetics of ion exchange in soil organic matter. III. Differential ion exchange reactions of Pb^{2+} ions in humic acid and peat. *J. Soil Sci.* 25:517–532.

Burden, D. S. 1989. Kinetics of Chromate and Phosphate Sorption by Oxide Minerals and Soils. PhD Dissertation. Louisiana State University, Baton Rouge, LA. p. 156.

Cameron, D. and A. Klute. 1977. Convective-disperisve solute transport with chemical equilibrium and kinetic adsorption model. *Water Resour. Res.* 13:183–188.

Carnahan, P., H. Luther, and J. O. Wilkes. 1969. *Applied Numerical Methods*, Wiley, New York.

Carski, T. H., and D. L. Sparks. 1985. A modified miscible displacement technique for investigating adsorption-desorption kinetics in soils. *Soil Sci. Soc. Am. J.* 49:1114–1116.

Carski, T. H., and D. L. Sparks. 1987. Differentiation of soil nitrogen fractions using a kinetic approach. *Soil Sci. Soc. Am. J.* 51:314–317.

Carslaw, H. S. and J. C. Jaeger. 1959. *Conduction of Heat in Solids*, Clarendon, Oxford, p. 492.

Cederberg, G. A., R. L. Street, and O. J. Leckie. 1985. A groundwater mass transport and equilibrium chemistry model for multicomponent systems. *Water Resour. Res.* 21:1095–1104.

Chaudhari, N. M. 1971. An improved numerical technique for solving multi-dimensional miscible displacement equations. *Soc. Petr. Eng. J.* 11:277–284.

Chen, C. and R. J. Wagenet. 1992. Simulation of water and chemicals in macropore soils. I. Representation of the equivalent macropore influence and its effect on soil water flow. *J. Hydrol.* 130:105–126.

Cho, C. M. 1971. Convective transport of ammonium with nitrification in soil. *Can. J. Soil Sci.* 51:339–350.

Chou, L. and R. Wollast. 1984. Study of the weathering of albite at room temperature and pressure with a fluidized bed reactor. *Geochim. Cosmochim. Acta* 48:2205–2217.

Cleary, R. W. and D. D. Adrian. 1973. Analytical solution of the convective-dispersive equation for cation adsorption, *Soil Sci. Soc. Am. Proc.*, 37:197–199.

Clothier, B. E., M. B. Kirkham, and J. E. Mclean. 1992. *In situ* measurement of the effective transport volume for solute moving through soil. *Soil Sci. Soc. Am. J.* 56:733–736.

Coats, K. H. and B. D. Smith. 1964. Dead-end pore volume and dispersion in porous media. *Soc. Pet. Eng.* 4:73–84.

Cooperband, L. R. and T. J. Logan. 1994. Measuring *in situ* changes in labile soil phosphorus with anion-exchange membranes. *Soil Sci. Soc. Am. J.* 58:105–114.

Danckwerts, O. V. 1953. Continuous flow systems. *Chem. Eng. Sci.* 2:1–13.

Daniel, C. and F. S. Wood. 1973. *Fitting Equations to Data*. Wiley-Interscience, New York.

Davidson, J. M. and R. K. Chang. 1972. Transport of picloram in relation to soil physical conditions and pore-water velocity, *Soil Sci. Soc. Am. Pro.*, 36:257–261.

Davidson, J. M., C. E. Rieck, and P. W. Santelmann. 1968. Influence of water flux and porous material on the movement of selected herbicides. *Soil Sci. Soc. Am. Proc.* 32:629–633.

Davis, J. A. and K. F. Hayes. 1986. *Geochemical Processes at Mineral Surfaces*. ACS Symp. Ser. 323. Am. Chem. Soc., Washington, D. C.

DeCamargo, O. A., J. W. Biggar, and D. R. Nielsen. 1979. Transport of inorganic phosphorus in an Alfisol. *Soil Sci. Soc. Am. J.* 43:884–890.

De Smedt, F. and P. J. Wierenga. 1979. Mass transfer in porous media with immobile water. *J. Hydrol.* 41:59–67.

De Smedt, F. and P. J. Wierenga. 1984. Solute transfer through columns of glass beads. *Water Resour. Res.* 20:225–232.

Dowdy, R. H. and V. V. Volk. 1983. Movement of heavy metals in soils in chemical mobility and reactivity in soil systems. Soil Science Society of America, Special Publication Number 11. Madison, WI, p. 229–240.

Dzombak, D. A., and F. M. M. Morel. 1990. *Surface Complexation Modeling. Hydrous Ferric Oxide.* John Wiley & Sons, New York.

Eary, L. E. and D. Rai. 1991. Chromate reduction by subsurface soils under acidic conditions. *Soil Sci. Soc. Am. J.* 55:676–683.

Ellis, B. G., B. D. Knezek, and L. W. Jacobs. 1983. The movement of micronutrients in soils. In: *Chemical Mobility and Reactivity in Soil Systems*, D. W. Nelson (Ed.). SSSA Spec. Publ. 11, ASA, Madison, WI, chap. 8, p. 109–122.

ElPrince, A. M. and G. Sposito. 1981. Thermodynamic derivation of equations of the Langmuir type for ion exchange equilibria in soils. *Soil Sci. Soc. Am. J.* 45:277–282.

Elrashidi, M. A. and G. A. O'Connor. 1982a. Boron sorption and desorption in soils. *Soil Sci. Soc. Am. J.* 46:27–31.

Elrashidi, M. A. and G. A. O'Connor. 1982b. Influence of solution composition on sorption of zinc by soils. *Soil Sci. Soc. Am. J.* 46:1153–1158.

Evans, R. L. and J. J. Jurinak. 1976. Kinetics of phosphate release from a desert soil. *Soil Sci.* 121:205–211.

Fiskell, J. G. A., R. S. Mansell, H. M. Selim, and F. G. Martin. 1979. Kinetic behavior of phosphate sorption by acid sandy soil. *J. Environ. Qual.* 8:579–584.

Flühler, H., J. Polomski, and P. Blaser. 1982. Retention and movement of fluoride in soils. *J. Environ. Qual.* 11:461–468.

Frost, A. A. and R. G. Pearson. 1961. *Kinetics and Mechanism.* Wiley, New York.

Fuller, W. H. 1977. Movement of selected metals, asbestos, and cyanide in soil: applications to waste disposal problems. EPA-600/2-77-020. U.S. EPA, Cincinnati, OH.

Fuller, C. C., J. A. Davis, and G. A. Waychunas. 1993. Surface chemistry of ferrihydrite. Kinetics of arsenate adsorption and coprecipitation. *Geochim. Cosmochim. Acta* 57:2271–2282.

Förster, U. and G. T. W. Wittman. 1981. *Metal Pollution in the Aquatic Environment.* Springer-Verlag, Berlin.

Gaines, G. L. and H. C. Thomas. 1953. Adsorption studies on clay minerals. II. A formulation of the thermodynamics of exchange adsorption. *J. Chem. Phys.* 21:714–718.

Garcia-Miragaya, J. and A. L. Page. 1976. Influence of ionic strength and inorganic complex formation on the sorption of trace amounts of Cd by montmorillonite. *Soil Sci. Soc. Am. Proc.* 40:658–663.

Gardiner, W. C., Jr. 1969. *Rates and Mechanisms of Chemical Reactions.* W. A. Benjamin, Menlo Park, CA.

Gaston, L. A. and H. M. Selim. 1990a. Transport of exchangeable cations in an aggregated clay soil. *Soil Sci. Soc. Am. J.* 54:31–38.

Gaston, L. A. and H. M. Selim. 1990b. Prediction of cation mobility in montmorillonitic media based on exchange selectivities of montmorillonite. *Soil Sci. Soc. Am. J.* 54:1525–1530.

Gaston, L. A. and H. M. Selim. 1991. Predicting cation mobility in kaolinite media based on exchange selectivities of kaolinite. *Soil Sci. Soc. Am. J.* 55:1255–1261.

Gerritse, R. G. and R. Singh. 1988. The relationship between pore water velocity and longitudinal dispersivity of Cl, Br, and D_2O in soils. *J. Hydrol.* 104:173–180.

Goldberg, S. and G. Sposito. 1984. A chemical model of phosphate adsorption by soils, I. Reference oxide minerals. *Soil Sci. Soc. Am. J.* 48:772–778.

Goltz, M. N. and P. V. Roberts. 1986. Three-dimensional simulations for solute transport in an infinite medium with mobile-immobile zone. *Water Resour. Res.* 22:1139–1148.

Goltz, M. N. and P. V. Roberts. 1988. Simulations of physical nonequilibrium solute transport models: application to a large scale field experiment. *J. Contamin. Hydrol.* 3:37–63.

Griffin, R. A. and R. G. Burau. 1974. Kinetic and equilibrium studies of boron desorption from soil. *Soil Sci. Soc. Am. Proc.* 38:892–897.

Grolimmund, D., M. Borkovec, P. Federer, and H. Sticher. 1995. Measurements of sorption isotherms with flow-through reactors. *Environ. Sci. Technol.* 29:2317–2321.

Gupta, S. P. and R. A. Greenkorn. 1973. Dispersion during flow in porous media with bilinear adsorption, *Water Resour. Res.* 5:1357–1368.

Hachiya, K., M. Ashida, M. Sasaki, H. Kan, T. Inoue, and T. Yasunaga. 1979. Study of the kinetics of adsorption-desorption of Pb^{2+} on a γ-Al_2O_3 surface by means of relaxation techniques. *J. Phys. Chem.* 83:1866–1871.

Harmsen, K. 1977. *Behavior of Heavy Metals in Soils.* Centre for Agriculture Publishing and Documentation, Wageningen, The Netherlands.

Harter, R. D. 1984. Kinetics of metal retention by soils: some practical and theoretical considerations. *Agron. Abstr.* p. 177.

Harter, R. D. and R. G. Lehmann. 1983. Use of kinetics for the study of exchange reactions in soils. *Soil Sci. Soc. Am. J.* 47:666–669.

Harter, R. D. 1989. A new modeling-compatible solution to the first-order kinetics equation. *Soil Sci.* 147:97–102.

Havlin, J. L., D. G. Westfall, and S. R. Olsen. 1985. Mathematical models for potassium release kinetics in calcareous soils. *Soil Sci. Soc. Am. J.* 49:371–376.

Hayes, K. F. and J. O. Leckie. 1986. Mechanism of lead ion adsorption at the goethite-water interface. In: *Geochemical Processes at Mineral Surfaces.* J. A. Davis and K. F. Hayes (Eds.) ACS Symp. Ser. 323:114–141. American Chemical Society, Washington, D.C.

Helfferich, F. G. 1962. *Ion Exchange.* McGraw-Hill, New York.

Henrici, P. 1962. *Discrete Variable Methods in Ordinary Differential Equations.* John Wiley & Sons, New York.

Hinz, C., B. Buchter, and H. M. Selim. 1992 . Heavy metal retention in soils: application of multisite models to zinc sorption. In: *Engineering Aspects of Metal-Waste Management.* I. K. Iskandar and H. M. Selim (Eds.) CRC/Lewis, Boca Raton, FL. p. 141–170.

Hinz, C, H. M. Selim, and L. A. Gaston. 1994. Effect of sorption isotherm type on predictions of solute mobility in soil. *Water Resour. Res.* 30:3013–3021.

Hinz, C. and H. M. Selim. 1994. Transport of Zn and Cd in soils: experimental evidence and modelling approaches. *Soil Sci. Soc. Am. J.* 58:1316–1327.

Hodges, S. C. and G. Johnson. 1987. Kinetics of sulfate adsorption and desorption by Cecil soil using miscible displacement. *Soil Sci. Soc. Am. J.* 51:323–331.

Holford, I. C. R. and G. E. G. Mattingly. 1975. The high- and low-energy phosphate adsorption surfaces in calcareous soils. *J. Soil Sci.* 26:407–417.

Holford, I. C. R., R. W. M. Wedderburn, and G. E. G. Mattingly. 1974. A Langmuir two-surface equation as a model of phosphate adsorption by soils. *J. Soil Sci.* 25:242–254.

James, B. R. and R. J. Bartlett. 1983. Behavior of chromium in soils. VII. Adsorption and reduction of hexavalent forms. *J. Environ. Qual.* 12:177–181.

Jardine, P. M. and D. L. Sparks. 1984. Potassium-calcium exchange in a multireactive soil system. I. Kinetics. *Soil Sci. Soc. Am. J.* 48:39–45.

Jardine, P. M., L. W. Zelazny, and J. C. Parker. 1985a. Mechanisms of aluminum adsorption on clay minerals and peat. *Soil Sci. Soc. Am. J.* 49:862–867.

Jardine, P. M., J. C. Parker, and L. W. Zelazny. 1985b. Kinetics and mechanisms of aluminum adsorption on kaolinite using a two-site nonequilibrium transport model. *Soil Sci. Soc. Am. J.* 49:867–873.

Jaynes, D. B., R. S. Bowman, and R. C. Rice. 1988. Transport of a conservative tracer in a field under continuous flood irrigation. *Soil Sci. Soc. Am. J.* 52:618–624.

Jaynes, D. B., S. D. Logsdon, and R. Horton. 1995. Field method for measuring mobile/immobile water content and solute transfer rate coefficient. *Soil Sci. Soc. Am. J.* 59:352–356.

Jennings, A. A. 1987. Critical chemical reaction rates for multicomponent groundwater contamination models. *Water Resour. Res.* 23:1775–1784.

Jennings, A. A. and D. J. Kirkner. 1984. Instantaneous equilibrium approximation analysis. *J. Hydrol. Div. ASCE.* 110:1700–1717.

Jennings, A. A., D. J. Kirkner, and T. L. Theis. 1982. Multicomponent equilibrium chemistry in groundwater quality models. *Water Resour. Res.* 18:1089–1096.

Jopony, M. and S. D. Young. 1987. A constant potential titration method for studying the kinetics of Cu^{2+} desorption from soil and clay minerals. *J. Soil Sci.* 38:219–228.

Kabata-Pendias, A. and H. Pendias. 1984. *Trace Elements in Soils and Plants.* CRC Press, Boca Raton, FL, chap. 3, 11.

Kent, D. B., J. A. Davis, L. C. D. Anderson, and B. A. Rea. 1995. Transport of chromium and selenium in a pristine sand and gravel aquifer: Role of adsorption processes. *Water Resour. Res.* 31:1041–1050.

Kinniburgh, D. G. 1986. General purpose adsorption isotherms. *Environ. Sci. Technol.* 20:895–904.

Kinniburgh, D. G. and M. L. Jackson. 1981. Cation adsorption by hydrous metal oxides and clay. p. 91–160. In: *Adsorption of Inorganics at Solid-Liquid Interfaces.* M. A. Anderson and A. J. Rubin, (Eds.) Ann Arbor Science, Ann Arbor, MI.

Kinniburgh, D. G., J. A. Barker, and M. Whitefield. 1983. A comparison of some simple adsorption isotherms for describing divalent cation adsorption by ferrihydrite. *J. Colloid. Interf. Sci.* 95:370–384.

Kirda, C., D. R. Nielsen, and J. W. Biggar. 1973. Simultaneous transport of chloride and water during infiltration. *Soil Sci. Soc. Ame. Proc.* 37:339–345.

Kirkner, D. J., A. A. Jennings, and T. L. Theis. 1985. Multisolute mass transport with chemical interaction kinetics. *J. Hydrol.* 76:107–117.

Koch, S. and H. Flühler. 1993. Non-reactive solute transport with micropore diffusion in aggregated porous media determined by a flow-interruption method. *J. Contam. Hydrol.* 14:39–54.

Kreft, A. and A. Zuber. 1978. On the physical meaning of the dispersion equation and its solution for different initial and boundary conditions. *Chem. Eng. Sci.* 33:1471–1480.

Kressman, T. R. E. and J. A. Kitchener. 1949. Cation exchange with a synthetic phenolsulphonate resin. V. Kinetics. *Disc. Faraday Soc.* 7:90–103.

Laidler, K. J. 1987. *Chemical Kinetics*, 3rd ed. Harper and Row, New York.

Langmuir, I. 1918. The adsorption of gases on plane surfaces of glass, mica and platinum. *J. Am. Chem Soc.* 40:1361–1402.

Lasaga, A. C. 1981. Rate laws of chemical reactions. In: *Kinetics of Geochemical Processes.* A. C. Lasaga and R. J. Kirkpatrick (Eds.) Vol. 8. Reviews in mineralogy. Mineralogical Society of America, Washington, D.C. p. 1–68.

Lasaga, A. C. and R. J. Kirkpatrick (Eds.) 1981. *Kinetics of Geochemical Processes.* Vol. 8. Reviews in mineralogy. Mineralogical Society of America, Washington, D. C.

Lai, Sung-Ho and J. J. Jurinak. 1972. Cation adsorption in one dimensional flow through soils, A numerical solution. *Water Resour. Res.* 8:99–107.

Lapidus, L. and N. L. Amundson. 1952. Mathematics for adsorption in beds. VI. The effect of longitudinal diffusion in ion exchange and chromatographic column. *J. Phys. Chem.* 56:984–988.

Laryea, K. B., D. E. Elrick, and M. J. L. Robin. 1982. Hydrodynamic dispersion involving cationic adsorption during unsaturated, transient water flow in soil. *Soil Sci. Soc. Am. J.* 46:667–671.

Li, Y. and M. Ghodrati. 1994. Preferential transport of nitrate through soil columns containing root channels. *Soil Sci. Soc. Am. J.* 58:653–659.

Li, Y. and M. Ghodrati. 1995. Transport of nitrate in soils as affected by earthworm activities. *J. Environ. Qual.* 24:432–438.

Lindstrom, F. T., R. Haque, V. H. Freed, and L. Boersma. 1967. Theory on movement of some herbicides in soils, Linear diffusion and convection of chemicals in soils, *Environ. Sci. Technol.* 1:561–565.

Luxmoore, R. J. 1981. Micro-, meso-, and macroporosity of soil. *Soil Sci. Soc. Am. J.* 45:671–672.

Ma, L. and H. M. Selim. 1994. Tortuosity, mean residence time and deformation of tritium breakthroughs from uniform soil columns. *Soil Sci. Soc. Am. J.* 58:1076–1085.

Mansell, R. S., Bloom, S. A., Selim, H. M. and Rhue, R. D. 1988. Simulated transport of multiple cations in soil using variable selectivity coefficients. *Soil Sci. Soc. Am. J.* 52:1533–1540.

Mansell, R. S., H. M. Selim, P. Kanchanasut, J. M. Davidson, and J. G. A. Fiskell. 1977. Experimental and simulated transport of phosphorus through sandy soils. *Water Resour. Res.* 13:189–194.

Marquardt, D. W. 1963. An algorithm for least-squares estimation of non-linear parameters. *J. Soc. Ind. Appl. Math.* 11:431–441.

Mendoza, R. E. and N. J. Barrow. 1987. Characterizing the rate of reaction of some Argentinean soils with phosphate. *Soil Sci.* 143:105–112.

Miller, D. M., W. P. Miller, and M. E. Sumner. 1988. A continuously stirred tank reactor for solid/solute adsorption studies. *Agron. Abstr.* ASA, Madison, WI, p. 201.

Miller, D. M., M. E. Sumner, and W. P. Miller. 1989. A comparison of batch- and flow-generated anion adsorption isotherms. *Soil Sci. Soc. Am. J.* 53:373–380.

Miller, C. W. and L. V. Benson. 1983. Simulation of solute transport in a chemically reactive heterogeneous system. Model development and application. *Water Resour. Res.* 19:381–391.

Montero, J. P., J. O. Munoz, R. Abeliuk, and M. Vauclin. 1994. A solute transport model for the acid leaching of copper in soil columns. *Soil Sci. Soc. Am. J.* 58:678–686.

Munns, D. N., and R. L. Fox. 1976. The slow reaction which continues after phosphate adsorption: kinetics and equilibrium in some tropical soils. *Soil Sci. Soc. Am. J.* 40:46–51.

Murali, V., and L. A. G. Aylmore. 1983. Competitive adsorption during solute transport in soils. I. Mathematical models. *Soil Sci.* 135:143–150.

Nielsen, D. R., M. Th. van Genuchten, and J. M. Biggar. 1986. Water flow and transport processes in the unsaturated zone. *Water Resour. Res.* 22:89S–108S.

Nkedi-Kizza, P., J. M. Biggar, H. M. Selim, M. Th. van Genuchten, P. J. Wierenga, J. M. Davidson, and D. R. Nielsen. 1984. On the equivalence of two conceptual models for describing ion exchange during transport through an aggregated soil. *Water Resour. Res.* 20:1123–1130.

Nkedi-Kizza, P., J. W. Biggar, M. Th. van Genuchten, M. Th., Wierenga, P. J., Selim, H. M., Davidson, D. M. and Nielsen, D. R. 1983. Modeling tritium and chloride 36 transport through an aggregated oxisol. *Water Resour. Res.* 19:691–700.

Ogata, A. 1970. Theory of Dispersion in Granular Medium. U. S. Geological Survey Professional Paper No. 411-I.

Ogwada, R. A. and D. L. Sparks. 1986a. A critical evaluation on the use of kinetics for determining thermodynamics of ion exchange in soils. *Soil Sci. Soc. Am. J.* 50:300–305.

Ogwada, R. A. and D. L. Sparks. 1986b. Kinetics of ion exchange on clay minerals and soil. I. Evaluation of methods. *Soil Sci. Soc. Am. J.* 50:1158–1162.

Ogwada, R. A. and D. L. Sparks. 1986c. Kinetics of ion exchange on clay minerals and soil. II. Elucidation of rate-limiting steps. *Soil Sci. Soc. Am. J.* 50:1162–1166.

Onken, A. B. and R. L. Matheson. 1982. Dissolution rate of EDTA-extractable phosphate from soils. *Soil Sci. Soc. Am. J.* 46:276–279.

Ozisik, M. N. 1968. *Boundary-Value Problems of Heat Conduction.* International Textbooks, Scranton, PA.

Papelis, C., P. V. Roberts, and J. O. Leckie. 1995. Modeling the rate of cadmium and selenite adsorption on micro- and meso-porous transition aluminas. *Environ. Sci. Technol.* 29:1099–1108.

Parker, J. C. and M. Th. van Genuchten. 1984. Determining transport parameters from laboratory and field tracer experiments. *Va. Agric. Exp. Sta. Bull.* 84-3.

Parker, J. C. and P. M. Jardine. 1986. Effect of heterogeneous adsorption behavior on ion transport. *Water Resour. Res.* 22:1334–1340.

Parker, J. C. and A. J. Valocchi. 1986. Constraints on the validity of equilibrium and first-order kinetic transport model in structured soils. *Water Resour. Res.* 22:399–407.

Passioura, J. B. 1971. Hydrodynamic dispersion in aggregated media. 1. Theory. *Soil Sci.* 111:339–344.

Patrick, W. H., Jr., B. C. Williams, and J. T. Moraghan. 1973. A simple system for controlling redox potential and pH in soil suspensions. *Soil Sci. Soc. Am. Proc.* 37:331–332.

Pavlatou, A. and N. A. Polyzopoulos. 1988. The role of diffusion in the kinetics of phosphate desorption: the relevance of the Elovich equation. *J. Soil Sci.* 39:425–436.

Peek, D. C. and V. V. Volk. 1985. Fluoride sorption and desorption in soils. *Soil Sci. Soc. Am. J.* 49:583–586.

Phelan, P. J. and S. V. Mattigod. 1987. Kinetics of heterogeneously initiated precipitation of calcium phosphates. *Soil Sci. Soc. Am. J.* 51:336–341.

Pinder, G. F. and W. Gray. 1977. *Finite Element Simulation in Surface and Subsurface Hydrology,* Academic Press, New York.

Randle, K. and E. H. Hartmann. 1987. Applications of the continuous flow stirred cell (CFSC) technique to adsorption of zinc, cadmium and mercury on humic acids. *Geoderma.* 40:281–296.

Rao, P. S. C., J. M. Davidson, R. E. Jessup, and H. M. Selim. 1979. Evaluation of conceptual models for describing nonequilibrium adsorption-desorption of pesticide during steady-state flow in soils. *Soil Sci. Soc. Am. J.* 43:22–28.

Rao, P. S. C., R. E. Jessup, D. E. Ralston, J. M. Davidson, and D. P. Kilcrease. 1980a. Experimental and mathematical description of nonadsorbed solute transfer by diffusion in spherical aggregate. *Soil Sci. Soc. Am. J.* 44:684–688.

Rao, P. S. C., D. E. Ralston, R. E. Jessup, and J. M. Davidson. 1980b. Solute transport in aggregated porous media: theoretical and experimental evaluation. *Soil Sci. Soc. Am. J.* 44:1139–1146.

Rasmuson, A. and I. Neretienks. 1981. Migration of radionuclides in fissured rock. The influence of micropore diffusion and longitudinal dispersion. *J. Geophys. Res.* 86:3749–3758.

Remson, I., G. M. Hornberger, and F. J. Molz. 1971. *Numerical Methods in Subsurface Hydrology*, Wiley, New York.

Rimstidt, J. D. and W. D. Newcomb. 1993. Measurement and analysis of rate data: the rate of reaction of ferric iron with pyrite. *Geochim. Cosmochim. Acta.* 57:1919–1934.

Rubin, J. 1983. Transport of reactive solutes in porous media, relation between mathematical nature of problem formulation and chemical nature of reactions. *Water Resour. Res.* 19:1231–1252.

Rubin, J. and R. V. James. 1973. Dispersion-affected transport of reacting solution in saturated porous media, Galerkin method applied to equilibrium-controlled exchange in unidirectional steady water flow. *Water Resour. Res.* 9:1332–1356.

Sadusky, M. C., D. L. Sparks, M. R. Noll, and G. J. Hendricks. 1987. Kinetics and mechanisms of potassium release from sandy Middle Atlantic Coastal Plain soils. *Soil Sci. Soc. Am. J.* 51:1460–1465.

Sardin, M., D. Schweich, F. J. Leij, and M. Th. van Genchten. 1993. Modeling the nonequilibrium transport of linearly interacting solutes in porous media: a review. *Water Resour. Res.* 27:2287–2307.

Schmidt, H. W. and H. Sticher. 1986. Long-term trend analysis of heavy metal content and translocation in soils. *Geoderma.* 38:195–207.

Schnabel, R. R. and D. J. Fitting. 1988. Analysis of chemical data from dilute, dispersed, well-mixed flow-through systems. *Soil Sci. Soc. Am. J.* 52:1270–1273.

Selim, H. M. and R. S. Mansell. 1976. Analytical solution of the equation of reactive solutes through soils. *Water Resour. Res.* 12:528–532.

Selim, H. M., J. M. Davidson, and R. S. Mansell. 1976. Evaluation of a two-site adsorption-desorption model for describing solute transport in soils. In: *Proceedings of the Summer Computer Simulation Conference*, Washington, D.C. Simulation Councils, LaJolla, CA, p. 444–448.

Selim, H. M. 1981. Modeling kinetic behavior of cadmium interaction in soils. In: *Proceedings of the Summer Computer Simulation Conference*, Washington, D. C. Simulation Councils, Inc., LaJolla, CA, p. 385–390.

Selim, H. M. and I. K. Iskandar. 1981. Modeling nitrogen transport and transformations in soils. 1. Theoretical considerations. *Soil Sci.* 131:233–241.

Selim, H. M., R. Schulin, and H. Flühler. 1987. Transport and ion exchange of calcium and magnesium in an aggregated soil. *Soil Sci. Soc. Am. J.* 51:876–884.

Selim, H. M., and M. C. Amacher. 1988. A second-order kinetic approach for modeling solute retention and transport in soils. *Water Resources Res.* 24:2061–2075.

Selim, H. M. 1989. Prediction of contaminant retention and transport in soils using kinetic multireaction models. *Environ. Health Perspec.* 83:69–75.

Selim, H. M., M. C. Amacher, and I. K. Iskandar. 1989. Modeling the transport of chromium(VI) in soil columns. *Soil Sci. Soc. Am. J.* 53:996–1004.

Selim, H. M., M. C. Amacher, and I. K. Iskandar. 1990. Modeling the transport of heavy metals in soils. CRREL-Monograph 90-2, U.S. Army Corps of Engineers, p. 158.

Selim, H. M., B. Buchter, C. Hinz, and L. Ma. 1992. Modeling the transport and retention of cadmium in soils: multireaction and multicomponent approaches. *Soil Sci. Sci. Am. J.* 56:1004–1015.

Selim, H. M. 1992. Modeling the transport and retention of inorganics in soils. *Adv. Agron.*, 47:331–384.

Selim, H. M., and Ma, L. 1995. Transport of reactive solutes in soils: a modified two-region approach. *Soil Sci. Soc. Am. J.* 59:75–82.

Seyfried, M. S., D. L. Sparks, A. Bar-Tal, and S. Feigenbaum. 1989. Kinetics of calcium-magnesium exchange on soil using a stirred-flow reaction chamber. *Soil Sci. Soc. Am. J.* 53:406–410.

Sidle, R. C., L. T. Kardos, and M. Th. van Genuchten. 1977. Heavy metal transport model in a sludge treated soil. *J. Environ. Qual.* 6:438–443.

Sivasubramaniam, S. and O. Talibudeen. 1972. Potassium-aluminum exchange in acid soils. I. Kinetics. *J. Soil Sci.* 23:163–173.

Skopp, J. 1986. Analysis of time-dependent chemical processes in soils. *J. Env. Qual.* 15:205–213.

Skopp, J. and D. L. McCallister. 1986. Chemical kinetics from a thin disc flow system: Theory. *Soil Sci. Soc. Am. J.* 50:617–623.

Smiles, D. E. and J. R. Philip. 1978. Solute transport during absorption of water by soil: laboratory studies and their practical implications. *Soil Sci. Soc. Am. J.* 42:537–544.

Smiles, D. E., J. R. Philip, J. H. Knight, and D. E. Elrick. 1978. Hydrodynamic dispersion during absorption of water by soil. *Soil Sci. Soc. Am. J.* 42:229–234.

Smiles, D. E. and B. N. Gardiner. 1982. Hydrodynamic dispersion during unsteady, unsaturated water flow in a clay soil. *Soil Sci. Soc. Am. J.* 46:9–14.

Sparks, D. L. 1985. Kinetics of ionic reactions in clay minerals and soils. *Adv. Agron.* 38:231–266.

Sparks, D. L. 1986. Kinetics of reactions in pure and mixed systems. In: *Soil Physical Chemistry*. D. L. Sparks (Ed.) CRC Press, Boca Raton, FL, p. 83–178.

Sparks, D. L. 1989. *Kinetics of Soil Chemical Processes*. Academic Press, San Diego, CA.

Sparks, D. L. and P. M. Jardine, 1981. Thermodynamics of potassium exchange in soil using a kinetics approach. *Soil Sci. Soc. Am. J.* 45:1094–1099.

Sparks, D. L. and P. M. Jardine. 1984. Comparison of kinetic equations to describe K-Ca exchange in pure and in mixed systems. *Soil Sci.* 138:115–122.

Sparks, D. L. and J. E. Rechcigl. 1982. Comparison of batch and miscible displacement techniques to describe potassium adsorption kinetics in Delaware soils. *Soil Sci. Soc. Am. J.* 46:875–877.

Sparks, D. L. and D. L. Suarez. (Eds.) 1991. *Rates of Soil Chemical Processes*. SSSA Spec. Publ. No. 27, Soil Science Society of America, Madison, WI.

Sparks, D. L., L. W. Zelazny, and D. C. Martens. 1980. Kinetics of potassium desorption in soil using miscible displacement. *Soil Sci. Soc. Am. J.* 44:1205–1208.

Sparks, D. L. and P. C. Zhang. 1991. Relaxation methods for studying kinetics of soil chemical phenomena. In: *Rates of Soil Chemical Processes*. D. L. Sparks and D. L. Suarez (Eds.) SSSA Spec. Publ. No. 27. Soil Sci. Soc. Am., Madison, WI, p. 61–94.

Sposito, G. 1980. Derivation of the Freundlich equation for ion exchange reactions in soils. *Soil Sci. Soc. Am. J.* 44:652–654.

Sposito, G. 1981. *The Thermodynamics of Soil Solutions*. Oxford University Press, New York.

Sposito, G. 1982. On the use of the Langmuir equation in the interpretation of "adsorption" phenomena. II. The "two-surface" Langmuir equation. *Soil Sci. Soc. Am. J.* 46:1157–1152.

Sposito, G. 1984. *The Surface Chemistry of Soils.* Oxford University Press, New York.

Sposito, G. 1986. Distinguishing adsorption from surface precipitation. In: *Geochemical Processes at Mineral Surfaces.* J. A. Davis and K. F. Hayes (Eds.) ACS Symp. Ser. 323. American Chemical Society, Washington, D.C. p. 217–228.

Sposito, G. 1989. *The Chemistry of Soils.* Oxford University Press, New York.

Sposito, G. 1994. *Chemical Equilibria and Kinetics in Soils.* Oxford University Press, New York.

Sposito, G. and S. V. Mattigod. 1977. On the chemical foundation of sodium adsorption ratio. *Soil Sci. Soc. Am. J.* 41:323–329.

Starr, J. L. and J.-Y. Parlange. 1979. Dispersion in soil columns; the snow plow effect. *Soil Sci. Soc. Am. J.* 45:448–450.

Stone, A. T. 1987a. Microbial metabolites and the reductive dissolution of manganese oxides: oxalate and pyruvate. *Geochim. Cosmochim. Acta.* 51:919–925.

Stone, A. T. 1987b. Reductive dissolution of manganese(III/IV) oxides by substituted phenols. *Environ. Sci. Technol.* 21:979–988.

Stone, A. T. and J. J. Morgan. 1984. Reduction and dissolution of manganese(III) and manganese(IV) oxides by organics. 2. Survey of the reactivity of organics. *Environ. Sci. Technol.* 18:617–624.

Stumm, W. 1986. Coordinative interactions between soil solids and water—an aquatic chemist's point of view. *Geoderma,* 38:19–30.

Stumm, W. and J. J. Morgan. 1981. *Aquatic Chemistry.* John Wiley, New York.

Swartjes, F. A., M. Renger, and G. Wessolek. 1992. Solubility and numerical simulation of the dynamics of some heavy metals in a polluted soil—a case study. In: *Engineering Aspects of Metal-Waste Management.* I. K. Iskandar and H. M. Selim (Eds.) CRC/Lewis, Boca Raton, FL, p. 171–179.

Theis, T. L. 1988. Reactions and transport of trace metals in groundwater. In: *Metal Speciation: Theory, Analysis, and Application.* J. R. Kramer and H. E. Allen (Eds.) CRC/Lewis, Boca Raton, FL.

Theis, T. L., R. Iyer, and L. W. Kaul. 1988. Kinetic studies of cadmium and ferricyanide adsorption on goethite. *Environ. Sci. Technol.* 22:1013–1017.

Tiller, K. G., J. Gerth, and G. Brümmer. 1984. The relative affinities of Cd, Ni, and Zn for different soils clay fractions. Procedures and partitioning of bound forms and their interpretations. *Geoderma.* 34:1–16.

Tiller, K. G., V. K. Nayyar, and P. M. Clayton. 1979. Specific and nonspecific sorption of cadmium by soil clays as influenced by zinc and calcium. *Aust. J. Soil Res.* 17:17–28.

Toner, C. V., IV and D. L. Sparks. 1995. Chemical relaxaation and double layer model analysis of boron adsorption on alumina. *Soil Sci. Soc. Am. J.* 59:395–404.

Toner, C. V., IV, D. L. Sparks, and T. H. Carski. 1989. Anion exchange chemistry of middle Atlantic soils: charge properties and nitrate retention kinetics. *Soil Sci. Soc. Am. J.* 53:1061–1067.

Toride, N., F. J. Leij, and M. Th. van Genuchten. 1993. A comprehensive set of analytical solution for nonequilibrium solute transport with first-order decay and zero-order production. *Water Resour. Res.* 29:2167–2182.

Travis, C. C. and E. L. Etnier. 1981. A survey of sorption relationships for reactive solutes in soil. *J. Environ. Qual.* 10:8–17.

Valocchi, A. J. 1985. Validity of the local equilibrium assumption for modeling sorbing solute transport through homogeneous soils. *Water Resour. Res.* 21:808–820.

Valocchi, A. J. 1990. Use of temporal moment analysis to study reactive solute transport in aggregated porous media. *Geoderma.* 46:233–247.

Valocchi, A. J., R. J. Street, and P. V. Roberts. 1981. Transport of ion-exchange solutes in groundwater. Chromatographic theory and field simulations. *Water Resour. Res.* 17:1517–1527.

van der Zee, S. E. A. T. M., L. G. J. Fokkink, and W. H. van Riemsdijk. 1987. A new technique for the assessment of reversibly adsorbed phosphate. *Soil Sci. Soc. Am. J.* 51:599–604.

van der Zee, S., F. Leus, and M. Louer. 1989. Prediction of phosphate transport in small columns with an approximate sorption kinetics model. *Water Resour. Res.* 25:1353–1365.

van Genuchten, M. Th. 1981. Non-equilibrium transport parameters from miscible displacement experiments, Res. Report No. 119, U. S. Salinity Lab., Riverside, CA. p. 80.

van Genuchten, M. Th. 1985. A general approach for modeling solute transport in structured soils. *Proc. 17th Int. Congress. IAH, Hydrogeology of Rocks of Low Permeability.* Jan. 7–12, 1985, Tucson, AZ. Mem. Int. Assoc. Hydrogeol., 17:512–526.

van Genuchten, M. Th., J. M. Davidson, and P. J. Wierenga. 1974. An evaluation of kinetic and equilibrium equations for the prediction of pesticide movement in porous media. *Soil Sci. Soc. Am. Proc.* 38:29–35.

van Genuchten, M. Th. and W. J. Alves. 1982. Analytical solutions of the one-dimensional convective-dispersive solute transport equation. U. S. Department of Agriculture. Technical Bulletin No. 161. p.151.

van Genuchten, M. Th. and F. N. Dalton. 1986. Models for simulating salt movement in aggregated field soils. *Geoderma.* 38:165–183.

van Genuchten, M. Th. and J. C. Parker. 1984. Boundary conditions for displacement experiments through short laboratory soil columns. *Soil Sci. Soc. Am. J.* 40:473–480.

van Genuchten, M. Th. and R. J. Wagenet. 1989. Two-site/two-region models for pesticide transport and degradation: theoretical development and analytical solutions. *Soil Sci. Soc. Am. J.* 53:1303–1310.

van Genuchten, M. Th. and P. J. Wierenga. 1976. Mass transfer studies in sorbing porous media I. Analytical solutions. *Soil Sci. Soc. Am. J.* 40:473–480.

van Genuchten, M. Th. and P. J. Wierenga. 1977. Mass transfer studies in sorbing porous media. II. Experimental evaluation with tritium (^3H_2O). *Soil Sci. Soc. Am. J.* 41:272–277.

van Genuchten, M. Th. and P. J. Wierenga. 1986. Solute dispersion coefficients and dispersion factors. In: *Methods of Soil Analysis, Part 1* A. Klute (Ed.) Agronomy Monograph No. 9 (2nd ed.), ASA, Madison, WI, Chap. 44, p. 1025–2054.

van Genuchten, M. Th., P. J. Wierenga, and G. A. O'Connor. 1977. Mass transfer studies in sorbing porous media. III. Experimental evaluation with 2, 4, 5-T. *Soil Sci. Soc. Am. J.* 41:278–284.

van Eijkeren, J. C. M. and I. P. G. Loch. 1984. Transport of cation solutes in sorbing porous medium, *Water Resour. Res.* 20:714–718.

van Riemsdijk, W. H. and J. Lyklema. 1980a. The reaction of phosphate with aluminum hydroxide in relation with phosphate bonding in soils. *Coll. Sur.* I:33-44.

van Riemsdijk, W. H. and J. Lyklema. 1980b. Reaction of phosphate with gibbsite ($Al(OH)_3$) beyond the adsorption maximum. *J. Coll. Interface Sci.* 76:55–66.

van Riemsdijk, W. H. and F. A. M. de Haan. 1981. Reaction of orthophosphate with a sandy soil at constant supersaturation. *Soil Sci. Soc. Am. J.* 45:261–266.

van Riemsdijk, W. H. and A. M. A. van der Linden. 1984. Phosphate sorption by soils. II. Sorption measurement technique. *Soil Sci. Soc. Am. J.* 48:541–544.

van Schaik, J. C. and W. D. Kemper. 1966. Chloride diffusion in clay-water system. *Soil Sci. Soc. Am. Proc.* 30:22–25.

Villermaux, J. 1974. Deformation of chromatographic peaks under the influence of mass transfer phenomena. *J. Chromatographic Sci.* 12:822–831.

Warrick, A. W., J. W. Biggar, and D. R. Nielsen. 1971. Simultaneous solute and water transfer for an unsaturated soil. *Water Resour. Res.* 7:1216–1225.

Wilson, G. V., P. M. Jardine, and J. P. Gwo. 1992. Modeling the hydraulic properties of a multiregion soil. *Soil Sci. Soc. Am. J.* 56:1731–1737.

Wittbrodt, P. R. and C. D. Palmer. 1995. Reduction of Cr(VI) in the presence of excess soil fulvic acid. *Environ. Sci. Tech.* 29:255–263.

Yang, J. E., E. O. Skogley, S. J. Georgitis, B. E. Schaff, and A. H. Ferguson. 1991. The phytoavailability soil test: development and verification of theory. *Soil Sci. Soc. Am. J.* 55:1358–1365.

Yasuda, H., R. Berndtsson, A. Bahri, and K. Jinno. 1994. Plot-scale solute transport in a semiarid agricultural soil. *Soil Sci. Soc. Am. J.* 58:1052–1060.

Zachara, J. M., C. C. Ainsworth, C. E. Cowan, and C. T. Resch. 1989. Adsorption of chromate by subsurface soil horizons. *Soil Sci. Soc. Am. J.* 53:418–428.

Zachara, J. M., C. E. Cowan, R. L. Schmidt, and C. C. Ainsworth. 1988. Chromate adsorption on kaolinite. *Clays Clay Miner.* 36:317–326.

Zachara, J. M., D. C. Girvin, R. L. Schmidt, and C. T. Resch. 1987. Chromate adsorption on amorphous iron oxyhydroxide in presence of major groundwater ions. *Environ. Sci. Technol.* 21:589–594.

Zasoski, R. J. and R. G. Burau, 1978. A technique for studying the kinetics of adsorption in suspensions. *Soil Sci. Soc. Am. J.* 42:372–374.

Zhang, P. C. and D. L. Sparks. 1989. Kinetics and mechanisms of molybdate adsorption/desorption at the goethite/water interface using pressure-jump relaxation. *Soil Sci. Soc. Am. J.* 53:1028–1034.

Zhang, P. and D. L. Sparks. 1990a. Kinetics and mechanisms of sulfate adsorption/desorption on goethite using pressure-jump relaxation. *Soil Sci. Soc. Am. J.* 54:1266–1273.

Zhang, P. and D. L. Sparks. 1990b. Kinetics of selenate and selinite adsorption/desorption at the goethite/water interface. *Environ. Sci. Technol.* 24:1848–1856.

INDEX

A

Arrhenius equation, 48–49
Activation energy, 49
Activity coefficient, 159
Adsorption, 21, 43, 60
Aggregates
  spherical, 138
  rectangular, 139
  inter, 136, 140
  intra, 135, 140
Alligator soil, 6–9
Asymptotic behavior, 14
Arsenic (As), 8

B

Batch methods, 22–23
Bentonite, 92, 93
Binary homovalent exchange,
  160–161
Boundary and initial conditions,
  84–85
Bulk density of soil, 56, 77
Brenner exact solution, 86–87, 90

C

Cadmium (Cd), 1, 8
  retention kinetics, 55, 56

kinetics, 74
release, 25
breakthrough, 111, 171, 172
Calcium (Ca)
  exchange isotherms, 161, 162
  breakthrough, 173–174
Cation exchange capacity (CEC), 7,
  9, 160
Cecil soil, 6, 10, 56, 89, 92, 109
Clearly and Adrian exact solution,
  87–91
Chromium (Cr)
  isotherms, 68, 127
  parameter, 8, 9, 68, 128
  kinetics, 70–71, 129–131
  breakthrough curves, 107–109,
  132, 133
Cobalt (Co), 8
Competitive ion exchange, 157
Complexation
  inner-sphere, 17, 19
  outer-sphere, 19
  surface, 17
Continuity equation, 77–80
Convective–dispersive equation,
  82–91
Copper (Cu), 1
  retention isotherm, 10
  adsorption–desorption, 60
Crank–Nickolson method, 97

D

Damkohler number, 112
Darcy flux, 80, 83
Desorption
  reaction, 40
  methods, 33
Diffusion
  apparent, 80
  ion exchange, 61
  film, 28
  mixing, 26
  molecular, 81
  particle, 28
Dispersion (*D*)
  estimation, 91
  longitudinal, 81
  mechanical, 81, 92
  numerical, 98
  transverse, 81
Dispersivity, 82
Dothan soil, 92
Dynamic soil region, 135–136

E

Elementary reactions, 17
Elovich model, 5
Equal affinity, 173–174
Equilibrium
  models, 4, 5
  constant, 18
  quasi, 57
  ion-exchange, 159–161
Equivalent fraction, 159
Eustis soil, 92, 171
Exact solutions, 85, 94
Explicit–implicit solution, 97

F

Finite-difference approximation, 97
First-order
  reaction, 56
  kinetics, 61

Fluvic acid, 181
Fraction of sites, 136
Freundlich retention
  equation, 4, 5, 57
  general, 12
Freundlich sorption isotherms, 68

G

Gaines–Thomas selectivity coefficient, 159
Gaussian elimination, 99
General Freundlich isotherm, 12
General Langmuir–Freundlich, 12
Glass beads, 92, 140
Goethite, 177

H

Hg (mercury) 8, 73
Hydraulic conductivity, 83
Hydrodynamic dispersion, 81 (See mechanical dispersion)
Hysteresis, 58, 60

I

Initial conditions, 84
Ion exchange
  equilibrium, 159–161
  isotherms, 163, 165
  kinetics, 20, 61, 165, 175
  selectivity, 160
Immobile-water, 135, 136, 140
Inner-sphere
  complexation, 17, 19
  metal–ligand, 177
Irreversible
  first-order, 60, 181
  kinetics, 61
  mechanisms, 181–182
Isotherm equation, general, 11

K

Kaolinite, 92, 93
Kinetics methods
  batch, 22
  column, 38
  dilution, 38
  flow, 34, 39
  fluidized bed, 38
  in-situ, 30
  infinite-sink, 42
  mixing, 26
  radiotracer, 30, 38
  relaxation, 22
  stirred-flow, 36
  thin disk, 35
Kula soil, 75

L

L-curve, 10
Lafitte soil, 6, 9
Langmuir
  equation, 10
  kinetics, 116, 166
  two-site, 11, 116, 127
  two-surface, 12
Lead (Pb)
  sorption, 8
  retention isotherms, 7
Linear distribution coefficient K_d, 5
Linear
  isotherm, 57
  kinetic reaction, 56
  retention, 5
Lindstrom exact solution, 87

M

Mahan soil, 92
Magnesium (Mg) exchange isotherms, 161–162
Mass transfer, 61, 136, 138–140, 165
Maximum sorption capacity, 10, 113
McLaren soil, 10, 60

Mechanical dispersion, 81
Mercury (Hg), 8, 9
  kinetics, 74
  release, 76
Mixing techniques, 26
Mobile–immobile models, 135–138
Mobile water, 135, 136, 140
Molecular diffusion in water, 81, 138
Molokai soil, 75
Molybdenum (Mo), 8
Multicomponent models, 157
Multiple ion systems, 161
Multiple reactions, 61
Multireaction models, 63

N

Nickel (Ni), 8
Nonlinear retention, 57
  kinetics, 57–59
  equilibrium, 57
  transport, 95
Norwood soil, 56
Nonreactive solutes, 101
Numerical solution, 97
Numerical dispersion, 98

O

Olivier soil, 56, 68, 92, 108
Optimization, 46
Outer-sphere
  complexation, 19
  metal–ligand, 177

P

Parameter-optimization, 46–48
Peclet number, 86, 90–91
pH, 49, 179
Piston flow, 81, 91
Pore-volume, 88
Pore-water velocity, 31
Power mode, 5

R

Rate functions reactions
  effect of concentration, 48
  effect of temperature, 48
  effect of pH, 49
  effect of ionic strength, 49
  multireaction, 60
  second-order, 60
  single, 60
Rate laws, 18
Reactant concentrations, 48
Reaction
  elementary, 17
  overall, 17
  order, 62, 64, 67
Rectangular aggregate diameter, 140
Relaxation method, 22
Release reactions, 40 (See
  desorption)
Release, cumulative, 71, 76
Retardation factor, 84
Reversible reactions, 64, 65
Richards flow equation, 83
Rothmund–Kornfeld equation, 5,
  164

S

SADF, 12
Second-order
  approach, 141
  kinetics, 5, 118
  reactions, 115
  two-site, 113
Separation factor, 160
Sharkey soil, 56, 92
Sigmoidicity of isotherms, 5, 11
Single-reaction kinetics, 56–57
Sink/source, 60, 61, 65 (See also
  irreversible)
Site affinity, 11–12

Sodium (Na) exchange isotherms,
  161–162
Sorbed phase concentration, 159, 161
Sorption, maximum, 5, 116 (See
  also adsorption)
Sorption specific, 166, 167
Spherical aggregate diameter, 138
Spodosol, 6
Stagnant soil region, 135–136
Steady-state flow, 83
Stirred-flow method, 61
Surface complexation, 17
  equilibrium, 175
  kinetic, 178

T

Temperature, 48
Thermodynamic constant, 159
Thin-disk flow method, 35
Tortuosity, 81, 138
Tracer solutes, 91
Transport
  column, 49
  equation, 79–80
  length, 139
  mechanisms, 80
Tridiagonal matrix, 99
Tritium breakthrough results, 93
Two-region, 135–138
Two-site models, 63

V

Valency, 159
Vanadium (V), 8
Vanselow selectivity coefficient, 159
Variable selectivity, 163 (See ion
  exchange)
Variable charge surfaces, 176, 179
Velocity, water, 78–80

W

Water content, 5, 56
Water flow equation, 83
Water flow, steady, 83
Water suction, 140
Water uptake, 83
Webster soil, 107
Windsor soil, 56, 57, 68, 92, 108

Y

Yolo soil, 92

Z

Zinc (Zn), 8
   breakthrough, 173
   ion exchange, 165
   retention isotherms, 14–15

T - #0133 - 101024 - C0 - 234/156/12 [14] - CB - 9780873714730 - Gloss Lamination